101 SOLVED NUCLEAR ENGINEERING PROBLEMS

J. A. CAMARA, PE

PROFESSIONAL PUBLICATIONS, INC. • BELMONT, CA 94002

101 SOLVED NUCLEAR ENGINEERING PROBLEMS

Printed in the United States of America

Professional Publications, Inc.
1250 Fifth Avenue, Belmont, CA 94002
(650) 593-9119
www.ppi2pass.com

Current printing of this edition: 1

Library of Congress Cataloging-in-Publication Data

Camara, J. A., 1956–
101 solved nuclear engineering problems.
p. cm.
Includes bibliographical references (p.).
ISBN 1-888577-30-4 (pbk.)
1. Nuclear engineering--Problems, exercises, etc. 2. Nuclear engineering--Mathematics. I. Title: One hundred and one solved nuclear engineering problems. II. Title: One hundred one solved nuclear engineering problems.
TK9145.C315 1998
621.48'076--dc21 98-31318
CIP

TABLE OF CONTENTS

PREFACE/ACKNOWLEDGMENTS v
HOW TO USE THIS BOOK vii
NOMENCLATURE ix
HISTORY OF NUCLEAR POWER x

1 NUCLEAR POWER SYSTEMS 1
2 NUCLEAR FUEL MANAGEMENT 9
3 NUCLEAR RADIATION 19
4 NUCLEAR THEORY 27
5 NUCLEAR INSTRUMENTATION 35
6 NUCLEAR ISSUES 41
7 NUCLEAR POWER SYSTEMS: SOLUTIONS 43
8 NUCLEAR FUEL MANAGEMENT: SOLUTIONS 67
9 NUCLEAR RADIATION: SOLUTIONS 91
10 NUCLEAR THEORY: SOLUTIONS 109
11 NUCLEAR INSTRUMENTATION: SOLUTIONS 131
12 NUCLEAR ISSUES: SOLUTIONS 145

APPENDIX 1: NUCLEAR POWER SYSTEMS A-1
APPENDIX 2: NUCLEAR FUEL MANAGEMENT A-3
APPENDIX 3: NUCLEAR RADIATION A-5
APPENDIX 4: NUCLEAR THEORY A-7
APPENDIX 5: NUCLEAR INSTRUMENTATION A-9

REFERENCES R-1

PREFACE

This book in general, and the problems specifically, are meant to fill the need for a concise yet thorough review of nuclear engineering. Although written as a study guide for the nuclear professional engineer (PE) examination, I meant it as the beginning point for review in any of the many fields of nuclear engineering.

It is thorough in that all the subjects on the examination, which cover the broad sweep of nuclear engineering, are covered. Further, the solutions sections explain even the incorrect answers in detail, allowing more information to be presented. Additionally, the reference section lists texts that provide coverage of nuclear topics to any depth desired.

This book is concise in that only 101 different types of problems are presented. Nevertheless, I think with careful study you will find that the knowledge you seek in nuclear engineering can be ascertained from somewhere in these pages.

If you are studying for the PE examination, I recommend solving the problems in Chaps. 1–6 and checking the solutions in Chaps. 7–12. If a quick review, or a specific piece of information, is what you seek, using the solutions directly may be more helpful.

Diligent use of *101 Solved Nuclear Engineering Problems* can enhance your mental skills in nuclear engineering. I hope it serves this purpose well.

Should you find an error, I hope two events occur: first that you let me know, and second, that you learn something from the discovery; I know I will! I would be most appreciative of suggestions for improvement, questions, and recommendations for expansion so that any future editions or similar texts will more nearly meet your needs. Please contact me via PPI or at idrivesubs@aol.com.

J. A. CAMARA, PE

ACKNOWLEDGMENTS

It is with the utmost in gratitude that I thank Mr. Lindeburg for taking the chance on a new author; without that first step, I would still be wondering how to make all this happen. His able Acquisition Editors Ms. Peggy Yoon, who took me through my first tentative steps in the process, and Ms. Aline Magee, who kept it all on track and encouraged me every step of the way, were instrumental.

The knowledge expressed in this book is the result of countless hours spent with me by teachers, instructors, mentors, and friends from whom I learned so much. I thank them all for taking the time to instruct and guide me. And a special thanks to those friends who inspired me through their actions, words, and support.

One of those who inspired was retired Captain C. E. Ellis, USN. To me, he will always be "the Engineer" and the one whose standards are the benchmark of excellence.

Chief Petty Officer Todd Kievit's proofing efforts were helpful in preventing several errors. He is living proof that excellence exists in both the officer and enlisted ranks.

The ultimate thanks goes to my family. My wife, GeorgiaAnn, assisted in all aspects of the work, kept me focused on the goal, and long ago recognized the diamond that so needed polishing. My son, Jac, helped with drawings and Internet research. My daughter, Cassiopeia, assisted with some of the typing and was tremendously supportive during the entire process.

Finally, a thanks to my mom. For after all, without her, nothing would have been possible.

J. A. CAMARA, PE

HOW TO USE THIS BOOK

This book is meant as a compendium of the nuclear engineering field. This is admittedly an ambitious goal for a work of 101 problems of differing types. However, a thorough study of the problem statements and solutions, along with the supplemental equations, terms, diagrams, tables, recommended references, and even the incorrect solutions that are supplied, will expose you to many aspects of nuclear engineering, providing a broad overview of the field. Because of this, the book can serve a number of purposes, allowing tailored reviews of differing depths and focuses.

The problems themselves are stated in Chapters 1 through 6, and the solutions follow in Chapters 7 through 12. All the information required to solve the problem is provided in the problem statement, the information section preceding the problem, or the appendices. Special terms are defined in the nomenclature section (or in the problem itself). The data tables in the Appendices are abbreviated and not meant for general use; they are provided to eliminate the need to refer to other sources to solve the problems. Appendices 1 through 5 contain information relevant to Chapters 1 through 5, respectively. The Reference List is provided for those who want additional material for review.

In general, both SI and English Engineering System units are used, reflecting standard practice in each particular subject area. Symbology also varies to reflect what is done in practice.

NUCLEAR ENGINEERS

Nuclear engineers preparing for the nuclear professional engineer (PE) examination will find this book directly suited to their needs. Chapters 1 through 5 each cover the scope of one subject area on the PE exam. Chapter 6 covers current nuclear issues. The number of questions in each chapter reflects the approximate percentage of questions in that area on the exam. All examination areas are covered. The individual examination subject areas, along with a breakdown of the topics in that area, follow.

- Nuclear Power Systems: NSSS; BOP; thermal hydraulic applications; PRA; energy generation
- Nuclear Fuel and Waste Management: material balance; fuel composition and design; economic analysis; depletion and burnup; radioactive materials, handling, storage, and transportation; high- and low-level waste disposal; low-level waste treatment
- Nuclear Radiation: protection and shielding; radioactive material control and monitoring; dose assessment; environmental surveillance; regulatory compliance; decontamination
- Nuclear Theory: criticality; kinetics; neutronics; analysis of critical and subcritical systems; single and multiple group calculations; point kinetics; bare and reflected systems; effects of strong absorbers; reactivity calculations
- Nuclear Instrumentation: radiation detection; sensors; instrumentation and control; counting statistics; electronics of instruments

Study all the areas, using the problems with or without the solutions, depending on your preference. Should you find you are weak in an area, use the Reference List as a starting point for further study. Good luck!

MECHANICAL ENGINEERS

Mechanical engineers will find Chapter 1, Nuclear Power Systems, a useful review of heat transfer fundamentals. Chapter 2, Nuclear Fuel and Waste Management, provides the requisite background for the many mechanical calculations related to the storage and movement of radioactive material.

HEALTH PHYSICS PROFESSIONS

If you work in or are associated with the health physics field, you will find the chapters on nuclear radiation and nuclear theory an excellent review of the fundamentals, including the practical and theoretical aspects of shielding. Chapter 5, Nuclear Instrumentation, offers a review of the instruments used every day in this fascinating field.

NUCLEAR PLANT OPERATORS /TECHNICIANS

As a nuclear plant operator or technician, you will recognize that the problems in this book are written using civilian nuclear plant information. Regulation problems reference Nuclear Regulatory Commission guidelines. Further, the problems reflect what you need to know (or be able to understand with minimal study) in order to perform your job competently. The entire book can be used in training sessions designed to review all or some of general areas covered, with individual problems providing topics of discussion.

STUDENTS

For students of nuclear engineering, this book is a self-contained problem-solving guide; you can review a topic, then work the problems using the data tables in the Appendices—all you need is your calculator. It is also a convenient resource for those studying elements of nuclear energy and its use in the production of electricity. For military personnel, this book provides an unclassified reference with basic calculations covering the broad sweep of nuclear power. If more information is desired on a topic, the Reference List provides a starting point for further study.

MILITARY

As mentioned in the Students' section, this book provides an unclassified reference for those working in nuclear power or related fields. It should be particularly useful as part of a theoretical review prior to examinations. In addition, as all the data and terminology reflect that used in civilian plants, this text can serve as a transition tool for those leaving the military with the intention of working in the nuclear field. It will provide an introduction to civilian terminology and the parameters in individual problems can provide an appreciation for the differences between designs.

Whatever your reason for selecting this book, I hope it serves you well. Enjoy the adventure we call learning!

NOMENCLATURE

NUCLEAR ENGINEERING ABBREVIATIONS AND TERMS

ABBREVIATIONS

BOP: balance of plant
BWR: boiling-water reactor
DG: diesel generator
ECCS: emergency core-cooling system
EFPD: effective full power days
HEPA: high-efficiency particulate (filter)
HVAC: heating, ventilating, and air conditioning
I&C: instrumentation and control
I&M: inspection and maintenance
LOCA: loss-of-coolant accident
NSSS: nuclear steam supply system
PRA: probability risk assessment
PWR: pressurized-water reactor
RCS: reactor coolant system
SD: shutdown
SG: steam generator
SU: startup
TG: turbine generator

TERMS

actinides: The elements with atomic numbers from 90–103 inclusive.

anticipated operational occurrences: Events expected to occur one or more times in the life of a nuclear power plant.

backscatter: The scattering of radiation in a direction generally opposite the original direction.

biological half-life: The time required for a certain substance in a biological system to be reduced to one-half its original value.

blanket: A fertile material placed around a core for the purpose of breeding or conversion.

buckling, geometric: The first eigenvalue of the flux equation; it depends on the geometry of the core and it characterizes a bare critical core.

buckling, material: Characterizes the flux in a system of uniform composition; it depends on the materials used.

buildup flux: The flux that accounts for the interaction of radiation with a medium.

cadmium cutoff: The energy below which the detector response would not change if the cadmium cover were replaced with a material opaque to neutrons below said energy.

Cerenkov radiation: The radiation emitted by a charged particle moving through a material with a speed greater than the speed of light in that material. (This is the blue glow so many have heard of and commented on.)

conversion: The operation of a reactor in such a manner as to produce one fissile material while another is consumed. In nuclear fuel management, the process of changing U_3O_8 to UF_6.

critical: The condition where a self-sustaining chain reaction occurs. The condition where the effective multiplication factor is one. The condition where the rate of neutron production, excluding those neutron sources not dependent on the fission rate (i.e., source neutrons), is equal to the rate of neutron loss.

Doppler broadening: The observed broadening of a given neutron cross section resonance due to thermal motion in the material of concern.

excess reactivity: That amount of reactivity in a given reactor beyond that required to establish criticality.

kerma: The sum of ionized particles' kinetic energies per unit mass due to interactions with non-ionizing particles. Used often in conjunction with neutron exposure.

nuclear spin: The intrinsic angular momentum of a nucleus. Units are given by $\hbar$, which is Planck's constant divided by 2π.

reactivity: The measure of the deviation from criticality.

reflector: The material placed adjacent to the core and designed to scatter the escaping neutrons back into the core.

resonance integral: The integral of the product of the absorption cross section and the reciprocal of the neutron energy. Related to the probability of resonance absorption.

void coefficient: The derivative of the reactivity with respect to a void (i.e., the removal of a given material).

HISTORY OF NUCLEAR POWER

It was the early Greek philosophers who first determined for western civilization that the our planet was comprised of a very few elements or basic substances. Empedocles of Akrargas, about 430 BC, decided these elements were earth, air, water and fire. A century later Aristotle added the "aether" of the heavens.

About 450 BC, in an argument about whether matter could be subdivided indefinitely, Democritus decided it could not and named the smallest particles "atoms," meaning "nondivisible."

Alchemists of the medieval ages attached properties to various substances in an effort to categorize elements. Many, however, were sidetracked in the effort to change base metal into gold; so much so, in fact, that the name alchemist was abandoned in favor of chemist and alchemy became chemistry.

In 1661, Boyle laid down the modern criterion for an element: a basic substance that cannot be broken down to any simpler substance after it is isolated from a compound. Following this, much effort was expended to identify such substances. Cavendish showed that water was in fact hydrogen and oxygen, while Lavoisier proved that air was oxygen and nitrogen. The Greek "elements" were no more.

Boyle, Newton, and Dalton were all convinced that atoms did indeed exist. An Italian chemist named Amedeo Avogadro applied the atomic theory to gases and determined that equal volumes contained equal numbers of atoms. In the nineteenth century, atoms moved from the abstract to the tangible as atomic weights were worked out (first using oxygen as the standard) and then Brownian motion explained. Mendeleev placed order on the growing information regarding atoms with his periodic table. In the twentieth century we have at last "seen" the atom. A field ion microscope was used to strip positively charged ions from the tip of a fine needle and shoot them at a fluorescent screen, producing a five million-fold magnification that made the atoms visible as bright dots. In 1970 a scanning electron microscope was used to detect individual uranium and thorium atoms.

It is the twentieth century from which most would date "nuclear history." For though many still call nuclear energy "atomic," it is actually the interior of the atom, the nucleus, which is manipulated to release energy.

Early in the century, J. J. Thompson "discovered" the electron. Following this, the shell structure of the atom was defined and all of chemistry and the periodic table had a solid theoretical foundation. It is from this foundation that an understanding of the interior of the atom grew.

In the early 1900s radioactivity was discovered. Not all the "rays" identified were electrons, however. Some of these rays were positively charged and must have originated elsewhere. Rutherford determined the basic structure of the atom by firing alpha particles at a thin foil of metal. When some were scattered directly back and some went completely through, he surmised the atom must consist of a small central core surrounded by electrons. From here, things progressed rapidly, with Bohr developing his "liquid drop" model of the nucleus in 1936.

To explore the nucleus, Fermi used neutrons because they were not repulsed by the protons' positive charged. In one of his experiments, he found new radioactive substances with properties he could not explain. The answer came from Austrian physicist Meitner working in Berlin: the uranium had undergone fission. The information was carried by Bohr to a conference in Washington. A letter was written by Einstein, at the insistence of Wigner and Teller, to President Franklin Delano Roosevelt, and the Manhattan project was born.

The first reactor was built at the University of Chicago in a "pile" under the football stadium and went critical on December 2, 1942. In 1954, the submarine USS Nautilus was launched, showing the usefulness of nuclear energy. The first US commercial plant began operation in 1958 in Shippingport, Pennsylvania. In 1965, a satellite powered by a small reactor was launched.

From the seas to space, this environmentally clean source of some 3000 years' worth of energy is the key to the future.

1 NUCLEAR POWER SYSTEMS

Nuclear Power Systems (NPS) problems 1–10 are based on the following information and illustration.

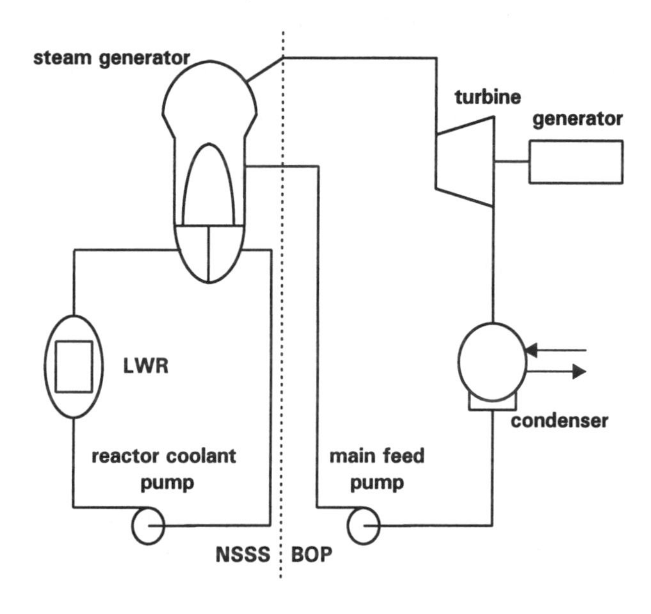

Terminology
BOP = balance of the plant
LWR = light water reactor
NSSS = nuclear steam supply system

Plant Data
active core height = 3.5 m
total fuel loading = 90 000 kg
UO_2 density = 10 g/cm^3
enrichment = 3 w/o [weight percent]
150 fuel assemblies
200 fuel pins per assembly
electrical generating capacity = 1000 MW
plant thermal efficiency (η_{th}) = 30%
plant availability = 75%

Scientific Data
E_r = energy recoverable per fission = 200 MeV
E_d = energy deposited in fuel per fission = 2.75×10^{-11} J
$\bar{\sigma}_{f,25}$ = average microscopic cross section for fission in U-235 = 366 barns[1,2]
c_p = specific heat capacity of water at 20°C (68°F) = 4.187 J/g·°C (1.00 Btu/lbm-°F)

NUCLEAR POWER SYSTEMS–1

What is the approximate thermal power output (P_{th}) of the core?

(A) 300 MW
(B) 1000 MW
(C) 1300 MW
(D) 3300 MW

NUCLEAR POWER SYSTEMS–2

What is the atom density of U-235 (N_{25}) in the core?

(A) 5.9×10^{20} cm^{-3}
(B) 6.7×10^{20} cm^{-3}
(C) 6.8×10^{20} cm^{-3}
(D) 7.6×10^{20} cm^{-3}

[1] The value shown is obtained by using the cross section value for U-235 tabulated for a neutron velocity of 2200 m/sec (the thermal value); averaging by assuming a Maxwellian distribution of neutron velocities (i.e., dividing by 1.128); multiplying by a non-1/v correction factor which is temperature dependent and also routinely tabulated; and finally, compensating for temperature of the uranium fuel as the above neutron cross sections are most often expressed at a temperature of 20°C. Also note that the symbols for the cross section vary from text to text (e.g., σ, $\bar{\sigma}$, $\bar{\sigma}'$, or $\hat{\sigma}$) as well as others, depending upon which of the operations mentioned are complete and the preferred symbology. See the Nuclear Theory section for additional information.
[2] The unit barns is sometimes symbolized with the letter b.

NUCLEAR POWER SYSTEMS–3

What is the average macroscopic fission cross section for the U-235 in the core?

(A) 2.16×10^{-1} cm^{-1}
(B) 2.45×10^{-1} cm^{-1}
(C) 2.49×10^{-1} cm^{-1}
(D) 2.78×10^{-1} cm^{-1}

NUCLEAR POWER SYSTEMS–4

What is the average thermal neutron flux in the core at the beginning of the first cycle?[3]

(A) 8.81×10^{11} neutrons/cm^2·s
(B) 2.93×10^{13} neutrons/cm^2·s
(C) 4.72×10^{13} neutrons/cm^2·s
(D) 3.00×10^{16} neutrons/cm^2·s

NUCLEAR POWER SYSTEMS–5

The core is expected to operate at 80% of full power the majority of the time. (Assume the fuel is uniformly distributed throughout the core.) Given this power level, what is most nearly the average thermal power density, P_d, in the fuel?

(A) 88 W/cm^3
(B) 296 W/cm^3
(C) 370 W/cm^3
(D) 300×10^3 W/cm^3

NUCLEAR POWER SYSTEMS–6

What is the mass of U-235 in the core?

(A) 2310 kg
(B) 2340 kg
(C) 2350 kg
(D) 2380 kg

[3] The beginning of first cycle was specified since the atom density of uranium 235 decreases with core operation and the average thermal neutron flux increases. See the Nuclear Theory section for more information.

NUCLEAR POWER SYSTEMS–7

Assume that during operation an average thermal neutron flux of 3.00×10^{13} neutrons/cm^3·s exists. The cycle for this plant is to be one year of operation followed by refueling of one-third of the core. What is the *fluence* (Φ) during one cycle?[4]

(A) 9.45×10^{13} neutrons/cm^2
(B) 3.15×10^{20} neutrons/cm^2
(C) 4.73×10^{20} neutrons/cm^2
(D) 9.46×10^{20} neutrons/cm^2

NUCLEAR POWER SYSTEMS–8

What is the fuel pin average thermal lineal power density (P_l)?

(A) 0.87 kW$_{th}$/ft
(B) 2.90 kW$_{th}$/ft
(C) 3.77 kW$_{th}$/ft
(D) 9.66 kW$_{th}$/ft

NUCLEAR POWER SYSTEMS–9

A certain reactor's heat sink (i.e., the cooling water used in its condenser) is a large lake. State regulations require that any water discharge to the lake be no more than 2°F greater than the mean temperature of the lake. Given this restriction, what mass flow rate ($\dot{m}$) must exist in the condenser?

(A) 1.71×10^9 lbm/hr
(B) 3.98×10^9 lbm/hr
(C) 5.63×10^9 lbm/hr
(D) 6.83×10^9 lbm/hr

[4] This information would be of concern to materials engineers and others attempting to determine the radiation effects on various portions of the plant. See the Nuclear Radiation section for more information.

NUCLEAR POWER SYSTEMS–10

Certain *engineered safety features* of this plant require electric power to operate. They are designed such that they are able to operate on AC or DC power. *Probability risk assessment* (PRA) studies for this plant show a 10^{-3} probability of DC power (i.e., battery power) not being available at any time during plant operation and a 10^{-2} probability of a "failure to start on demand" by the AC diesel generators. Related reliability studies indicate the occurrence of a failure of off-site power once per calendar year.[5] Further, plant emergency systems are required once in every ten of these failures.

What is the probability this plant's emergency systems will not operate due to a lack of electric power in any given year?

(A) 0.075×10^{-5}
(B) 0.100×10^{-5}
(C) 1.000×10^{-5}
(D) 1.100×10^{-2}

NUCLEAR POWER SYSTEMS–11

What is the *burnup rate* (BR) of a 500 MW_e U-235 fueled reactor with a recoverable energy per fission (E_r) of 200 MeV and a plant efficiency (η_{th}) of 25%?

(A) 2 g/day
(B) 200 g/day
(C) 525 g/day
(D) 2100 g/day

NUCLEAR POWER SYSTEMS–12

A steam generator is designed to support a steam mass flow rate ($\dot{m}$) of 122×10^6 lbm/hr with a maximum moisture content of 0.25%. The operating pressure is 1100 psia (P_{SG}) at a design feedwater inlet temperature (T_{FW}) of 440°F. What is the thermal power rating of the steam generator?

(A) 8.0×10^{10} Btu/hr
(B) 9.4×10^{10} Btu/hr
(C) 9.5×10^{10} Btu/hr
(D) 10.0×10^{10} Btu/hr

[5] Off-site power is the normal power supply to emergency systems.

NUCLEAR POWER SYSTEMS–13

A 1200 MW_e *pressurized water reactor* (PWR) with an overall thermal efficiency (η_{th}) of 30% is designed to supply saturated steam at a pressure of 1200 psia to two steam turbines. The exhaust of each turbine is cooled to near saturated liquid conditions at 4 psia in a single condenser. Assuming this PWR operates on an ideal Rankine cycle, calculate the thermal power transferred by the condenser.

(A) 9.05×10^3 Btu/hr
(B) 2.65×10^9 Btu/hr
(C) 9.05×10^9 Btu/hr
(D) 2.65×10^{15} Btu/hr

Nuclear Power Systems (NPS) problems 14–16 are based on the following information and illustration.

The rate of fission in a given fuel rod, and hence the rate of heat production, varies between fuel rods and is also a function of the position of the fuel rod in the core as given by the following expression.[6]

$$q'''(\vec{r}) = E_d \int_0^\infty \Sigma_{f,r}(E)\varphi(\vec{r},E)dE$$

q''' = power density, or power production per unit volume, as a function of position in the core[7]
E_d = energy deposited locally in the fuel per fission and normally considered to be ≈ 180 MeV
$\Sigma_{f,r}$ = macroscopic fission cross section of the fuel rod as a function of energy
φ = energy dependent flux as a function of radial position

Since, in a thermal reactor, most fissions are produced by thermal neutrons; and assuming a bare, finite cylinder as the reactor shape; and finally, calculating the total average macroscopic fission cross section for the core ($\Sigma_{f,25}$) to account for the rods ($\Sigma_{f,r}$) as well as all other components as if they were a homogeneous mixture; results in the following simplified expression.[8]

[6] The fuel rod is also referred to as a fuel pin, or simply pin. Rod when used alone may refer to the neutron absorber used to control the fission chain reaction. See the Nuclear Theory section for more information.
[7] The units depend on the desired units in the result; although, in most heat transfer calculations the English Engineering System continues to be used, hence Btu/hr-ft³.
[8] This cylindrical shape is used as the base assumption for many initial heat removal calculations.

$$q'''(r,z) = \left(\frac{1.16\,P\,E_d}{H\,a^2\,n\,E_r}\right) J_0\left(\frac{2.405r}{R}\right)\cos\left(\frac{\pi z}{H}\right)$$

P = total power
H = height of core [fuel rod height]
a = radius of the fuel rods
n = number of fuel rods
E_r = energy recoverable per fission ≈ 200 MeV
J_0 = Bessel function of the first kind of order zero
r = radial distance from the axis of the core
R = radius of the core
z = axial position along a given fuel rod usually represented as shown[9]

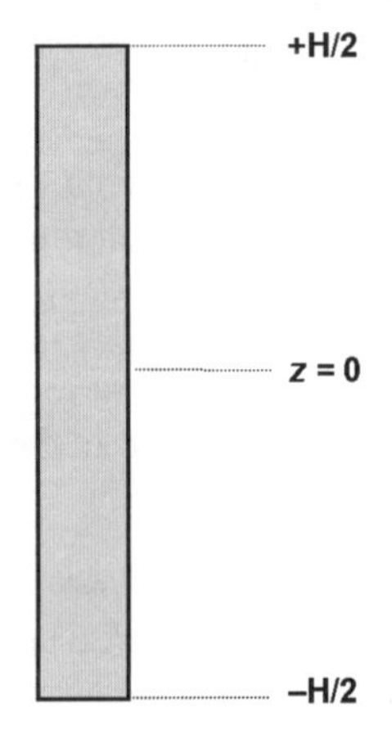

NUCLEAR POWER SYSTEMS–14

A 1000 MW_{th} research reactor is of cylindrical design, 6 ft in height and 8 ft in diameter. It contains 200 fuel assemblies each with 150 UO_2 cylindrical fuel rods with a diameter of 0.2 in. Assume the assemblies are uniformly distributed throughout the core and no *blanket* or *reflector* exists.[10] What is the power density in a fuel rod on the axis of the core, three-quarters of the distance from the bottom of the rod?

(A) 1.27×10^7 Btu/hr-ft^3
(B) 3.57×10^7 Btu/hr-ft^3
(C) 5.04×10^7 Btu/hr-ft^3
(D) 1.48×10^8 Btu/hr-ft^3

[9] Setting the zero point mid-axis eases calculations as will be shown in the associated problems.
[10] In other words, this is a bare reactor.

NUCLEAR POWER SYSTEMS–15

What is the power density at full power of a fuel rod located 2 ft from the axis, of the core described in the Prob. 14, at the point of its maximum power density?

(A) 1.2×10^7 Btu/hr-ft^3
(B) 4.8×10^7 Btu/hr-ft^3
(C) 5.8×10^7 Btu/hr-ft^3
(D) 1.4×10^8 Btu/hr-ft^3

NUCLEAR POWER SYSTEMS–16

What is the total power of the fuel rod located on the axis of the core when the core is operated at full power?

(A) 69×10^3 Btu/hr
(B) 85×10^3 Btu/hr
(C) 170×10^3 Btu/hr
(D) 240×10^3 Btu/hr

Nuclear Power Systems (NPS) problems 17 and 18 are based on the following information and illustrations.

A plate-type fuel element is shown from two views with the dimensions as indicated.

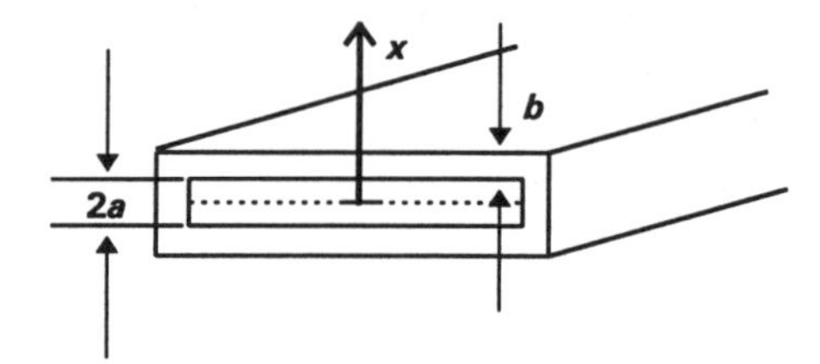

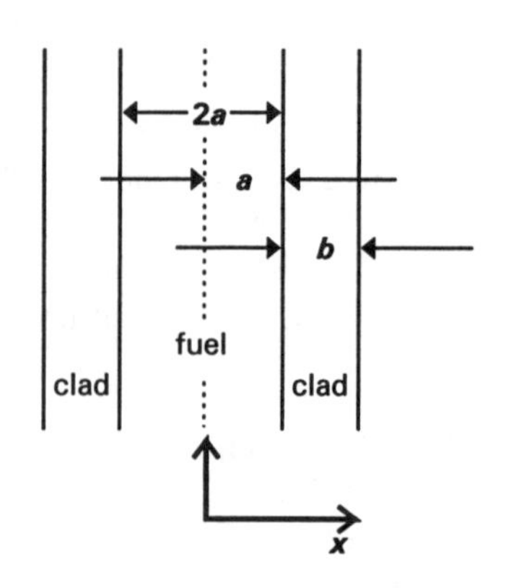

Plant Data
$2a = 0.40$ in
$b = 0.002$ ft
A = area of one face of the fuel plate = 4 ft^2

Scientific Data
thermal conductivity (k):
Zircaloy-4 $k_c = 5$ Btu/hr-ft-°F
UO_2 fuel $k_f = 3$ Btu/hr-ft-°F

film coefficient (h):
water $h_{H2O} = 2000$ Btu/hr-ft^2-°F

NUCLEAR POWER SYSTEMS–17

What is the total thermal resistance R from the center of the fuel to the coolant? (Ignore any resistance due to the bonding between the fuel and the cladding, which is normally very small.)

(A) 7.94×10^{-4} hr-°F/Btu
(B) 9.19×10^{-4} hr-°F/Btu
(C) 2.78×10^{-3} hr-°F/Btu
(D) 3.18×10^{-3} hr-°F/Btu

NUCLEAR POWER SYSTEMS–18

The temperature profile of this plate-type fuel element is as follows.

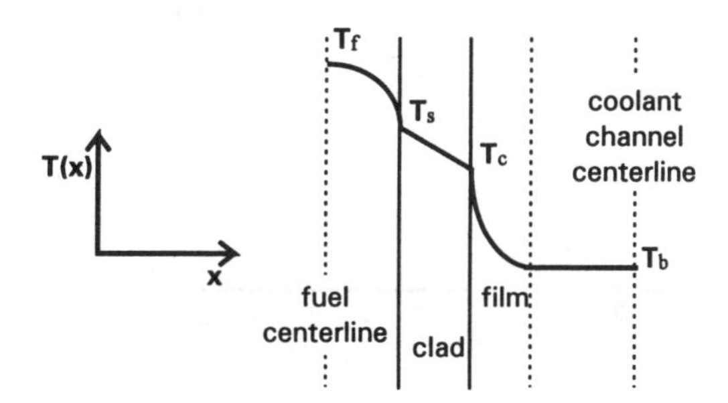

The ΔT between the bulk temperature (T_b) and the fuel temperature (T_f) at the centerline cannot exceed 100°F during steady state conditions, at the design T_b, in order to maintain the integrity of the fuel assembly.[11]

What is the limiting heat transfer rate for a single fuel plate?

(A) 32 kW
(B) 40 kW
(C) 48 kW
(D) 64 kW

NUCLEAR POWER SYSTEMS–19

A simplified system *fault tree* leading to a pipe rupture is shown below.

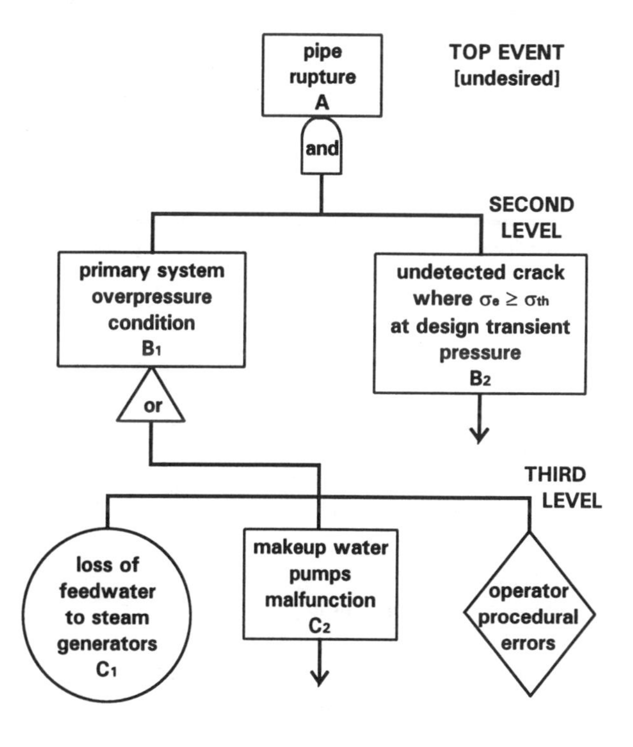

Symbology
circle = the probability of occurrence is known or can be calculated
rectangle = the probability must be traced from lower levels in the fault tree
diamond = the probability is not developed due to lack of information or the insignificance of the item
A, B, C = probability of the indicated event
σ_{th} = stress at the instant of fracture
σ_e = stress at the end of a crack

[11] That is, to keep the fuel inside the cladding. At the extreme, what this means is to avoid melting the fuel and its surrounding matrix support structure the temperature limits must be observed.

Using Boolean algebra, what is the probability of fault A?

(A) $(C_1 + C_2)B_2$
(B) $B_1 + B_2$
(C) $C_1 \cdot C_2 \cdot B_2$
(D) $C_1 + C_2 + B_2$

NUCLEAR POWER SYSTEMS–20

Consider a pipe rupture as an initiating event for a *loss-of-coolant accident* (LOCA). A reduced *event tree* showing the probabilities of the most likely consequences is shown. The probability of failure at each stage is labeled as P_1, P_2...with the probability of success (i.e., the system is available and functioning normally) represented by $1-P_2$, $1-P_3$, and so on. Radioactive release events are labeled I, II, III, and IV.

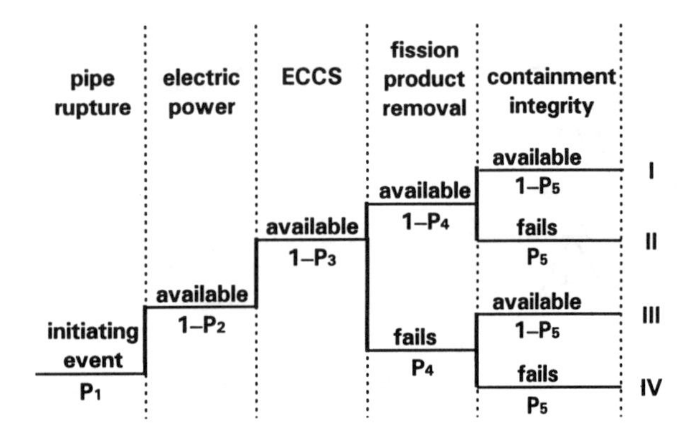

Terminology
ECCS = emergency core-cooling system

What is the probability of the most severe event (i.e., radioactive release I, II, III, or IV) shown?

(A) P_1
(B) $(P_1)(P_5)$
(C) $(P_1)(1-P_2)(1-P_3)(P_4)(1-P_5)$
(D) $(P_1)(P_4)(P_5)$

NUCLEAR POWER SYSTEMS–21

A pressurized water reactor (PWR) has a primary coolant inventory of 300 000 kg at 15.5 MPa_a and 315°C. the containment structure's free volume is 55 000 m^3. What is the pressure load on the containment building following a loss-of-coolant accident (LOCA)?

Consider this calculation a scoping calculation to determine an approximate value prior to an in-depth design calculation. Therefore, the thermal energy absorbed by plant components, the containment structure itself, and any injected water as well as heat input by fission product decay are to be ignored.

(A) 100 kPa_a
(B) 200 kPa_a
(C) 400 kPa_a
(D) 500 kPa_a

Nuclear Power Systems (NPS) problems 22–24 are based on the following information and illustration.

The following stress-strain curve is typical of that for common stainless steels (adjusted for normal power plant operating temperatures).[12]

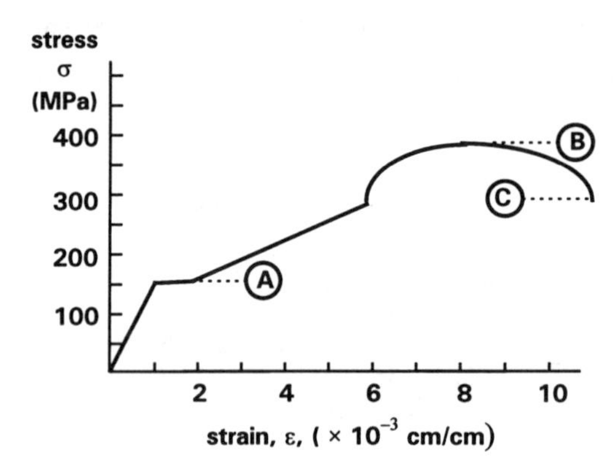

[12] This is an engineering or nominal stress-strain curve and is based on the original dimensions of the metal. True stress and strain depend on the actual dimensions at a given instant. See the materials portions of the various texts in the Reference section for additional information.

NUCLEAR POWER SYSTEMS–22

Point A represents which of the following?

(A) nil-ductility transition (NDT)
(B) ultimate tensile strength (UTS)
(C) work hardening point
(D) yield point (YP)

NUCLEAR POWER SYSTEMS–23

What is the value of Young's modulus?

(A) 1.50×10^5 Pa
(B) 7.50×10^{10} Pa
(C) 1.50×10^{11} Pa
(D) 4.00×10^{11} Pa

NUCLEAR POWER SYSTEMS–24

A certain reactor support assembly is constructed of steel and constrained along its x-axis as shown.

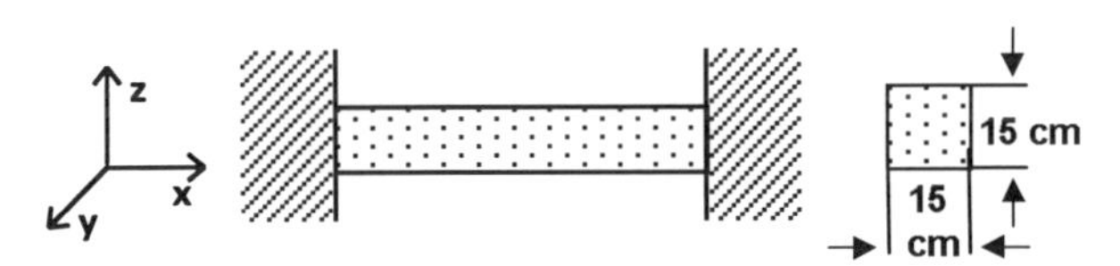

The assembly changes temperature from 20°C to 315°C during a startup from cold conditions. From a materials handbook, the coefficient of linear thermal expansion is 17×10^{-6} K^{-1}.

What is the compressive thermal stress on the assembly during hot operation?

(A) 3.8×10^8 Pa
(B) 7.5×10^8 Pa
(C) 2.0×10^9 Pa
(D) 1.7×10^{11} Pa

NUCLEAR POWER SYSTEMS–25

A simplified diagram of a natural circulation boiling-water reactor (BWR) is shown.

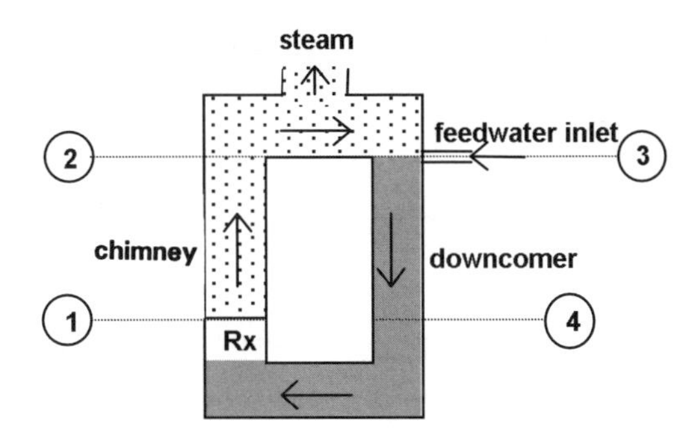

[steam water mixture]

[water]

Plant Data
chimney height = 10 m
steam pressure = 7.0 MPa
steam exit quality = 13%

The following three assumptions are used for this approximate calculation: (1) friction losses outside the core are negligible, (2) the system is isothermal, and (3) no heat transfer takes place to or from the surrounding environment.[13]

What is the approximate driving pressure available for natural circulation?[14]

(A) 70 kPa
(B) 75 kPa
(C) 80 kPa
(D) 85 kPa

[13] Clearly the isothermal condition does not occur. Additional data are required for a more accurate analysis. The incoming feed water will not be at the temperature of a saturated liquid regardless of how complete the mixing in the downcomer region.
[14] This is similar to the type of calculation necessary to determine the circulation head in a steam generator of a PWR plant.

NUCLEAR POWER SYSTEMS–26

Reactor control (i.e., the method of changing the neutron flux) involves the use of control rods. Additionally, both fixed burnable poisons and those in solution in the primary coolant are used to control the flux. A table of three common materials used in such applications and their properties follows.

designation	material
I	boron (B)
II	hafnium (Hf)
III	cadmium (Cd)

Properties

A—fairly large neutron absorption cross section over multiple isotopes

B—neutron absorption cross section in the thousands of barns

C—low melting point

D—resistant to corrosion in high temperature water

E—utilized in primary systems' coolant as a chemical shim

F—capable of decreasing reactor power level rapidly in its normal form as a control rod

Match the material with its properties. (The material may have more than one property associated with it and the properties may be used more than once.)

(A) I AC II BAC III BCD
(B) I AF II BCD III BCE
(C) I BF II AEF III BDE
(D) I BE II ADF III BCF

2 NUCLEAR FUEL MANAGEMENT

Nuclear Fuel Management (NFM) problems 1–5 are based on the following information and equations.

The uranium required as fuel for PWRs is mined as U_3O_8, purified using solvent extraction processes to remove impurities,[1] and then converted using chemical processes to uranium hexafluoride (UF_6).[2] The UF_6 is then enriched, usually to a nominal value of 3 w/o [weight percent] of U-235, using a gaseous diffusion process.[3]

The gaseous diffusion process is explained by two major equations.

$$F = P + W$$

$$x_f F = x_p P + x_w W$$

x_f = weight fraction of U-235 in the feed material[4]
x_p = weight fraction of U-235 in the product (this is the desired enrichment)
x_w = weight fraction of U-235 in the waste stream (this is the depleted uranium); also called the tails assay
F = number of kilograms of feed material, per unit time
P = number of kilograms of product enriched, per unit time
W = number of kilograms of uranium in the waste stream, per unit time[5]

[1] In the U.S., the PUREX process or a method using uranium peroxide ($UO_4 \cdot 2H_2O$) is utilized.
[2] The process used is dry hydroflour or wet solvent extraction.
[3] Other methods possible include the centrifuge, separation nozzle and ALVIS (Atomic Vapor Laser Isotope Separation).
[4] Note that this is a fraction, not a percent. Thus, natural uranium, at present, consists of 0.711% U-235. Therefore, x_f is 0.00711.
[5] F, P, and W are defined as mass flow rate (mass/time). These terms are frequently referred to as the "total" mass, without reference to time. No error is introduced during calculations if one is consistent.

One quantity derived from the gaseous diffusion process equations is the *feed factor*.

$$\frac{F}{P} = \frac{x_p - x_w}{x_f - x_w}$$

Another quantity derived from these equations is the *waste factor*.

$$\frac{W}{P} = \frac{x_p - x_f}{x_f - x_w} = \frac{F}{P} - 1$$

NUCLEAR FUEL MANAGEMENT–1

The feed factor F/P is a tabulated quantity which indicates the number of kilograms of uranium needed as feed for the enrichment process per kilogram of product. If a utility specifies an enrichment x_p of 3% with a tails assay of 0.3%, what is the feed factor?

(A) 0.04
(B) 0.40
(C) 3.30
(D) 6.60

NUCLEAR FUEL MANAGEMENT–2

During annual refueling a batch of fuel assemblies consisting of one-third of the 33 000 kg of 2.7 w/o (i.e., weight percent) U-235 will be replaced.

The cost of natural uranium is $40/kg. The cost of *conversion* from U_3O_8 to UF_6 is $5/kg and suffers a material loss of 0.5% uranium. Ignore any fabrication losses. The tails assay for the replacement fuel is 0.3%.

What is the cost through conversion of the natural uranium that must be provided as feed to an enrichment plant to provide for the year's refueling needs?

(A) $65,000
(B) $264,000
(C) $2,900,000
(D) $8,700,000

NUCLEAR FUEL MANAGEMENT–3

A separative work unit (SWU) is a quantity directly related to the resources required to perform the enrichment to the desired level of x_p, given values of x_f and x_w, and is given by

$$\text{SWU} = \left(PV(x_p) + WV(x_w) - FV(x_f)\right)T$$

SWU = separative work unit in kg or kg-SWU[6]
$V(x_i)$ = separation potentials with

$$V(x_i) = (2x_i - 1)\ln\left(\frac{x_i}{1-x_i}\right)$$

T = time period, usually a year

Assume the nominal one year time frame. What is the number of kilograms of natural uranium that must be provided as feed to the enrichment plant if one requests 100 000 kg[7] at 2.8% in U-235 enriched uranium with a known tails assay of 0.25%? Also, what is the number of SWUs needed for the separation?

(A) 100×10^3 kg; 3×10^6 kg-SWU
(B) 103×10^3 kg; 3×10^6 kg-SWU
(C) 48×10^3 kg; 340×10^3 kg-SWU
(D) 553×10^3 kg; 340×10^3 kg-SWU

NUCLEAR FUEL MANAGEMENT–4

While the SWU provides information on the resources required to enrich the desired amount of product, it varies with the amount of product and the time frame. To standardize, a SWU factor S is defined as the number of SWUs per unit of product.[8] In equation form,

$$S = \frac{\text{SWU}}{PT} = V(x_p) + \left(\frac{W}{P}\right)V(x_w) - \left(\frac{F}{P}\right)V(x_f)$$

P is the number of kilograms per unit time and T is the time, usually one year.[9]

What is the SWU factor for a 3% enrichment of natural uranium using a 0.25% tails assay?

(A) 2.715
(B) 3.780
(C) 3.811
(D) 30.588

[6] The SWU portion of the unit shown is not a unit per se. It's used for convenience in calculating the separative work unit factor S as will be shown in a later problem.

[7] This refers to the kg of enriched product (UF_6), not just uranium.
[8] The factor S is technically dimensionless. Its units are kg-SWU/kg. The SWU factor S provides the number of SWUs required per kilogram of product, not per gram or lbm.
[9] The SWU factor S is usually tabulated along with values of the feed factor F/P.

NUCLEAR FUEL MANAGEMENT–5

The following information is taken from the technical documentation provided by an enrichment facility:

P_u = price of natural uranium = $50/kg
P_c = price of conversion = $8/kg
P_{SWU} = price of separative work units = $90/SWU
l_c = conversion loss = 0.4%
x_w = tails assay = 0.3%

A plant requires 2.7 w/o (i.e., weight percent) enriched uranium for an upcoming refueling. What is the price per kilogram of enriched uranium required by the plant? Ignore fabrication losses.

(A) $300/kg
(B) $600/kg
(C) $800/kg
(D) $1100/kg

Nuclear Fuel Management (NFM) problems 6 and 7 are based on the following information and illustration.

A common fabrication process for pellet type fuel is one which starts with enriched uranium hexafluoride (UF_6) supplied in a high pressure cylinder as a solid. It is then heated, causing the UF_6 to sublime. The gas bubbles through water, forms UO_2F_2, and is mixed with ammonia water to precipitate the uranium as ammonium diuranate $(NH_4)_2U_2O_2$. The precipitate is dried at high temperature (*calcined*) to form U_3O_8. Hydrogen is used to further reduce the compound to uranium dioxide, UO_2. This is ground into a fine powder added to an adhesive agent and then pressed into a cylindrical pellet. The pellets are then sintered near the melting point to cause densification to occur. The pellets are then loaded into a fuel rod assembly such as shown.

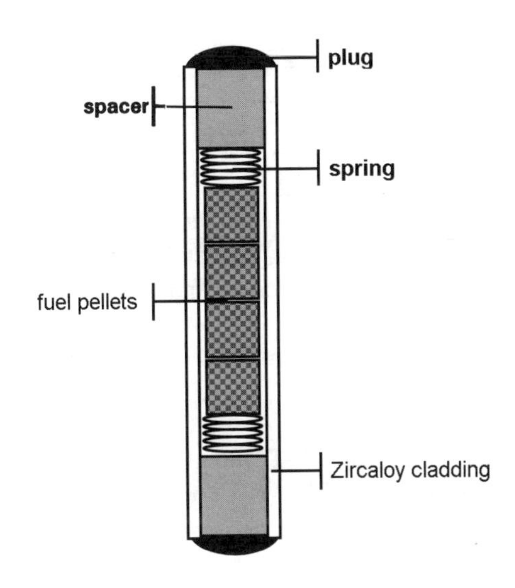

NUCLEAR FUEL MANAGEMENT–6

Which of the following is not a reason for the use of uranium dioxide (UO_2) as a form of nuclear fuel?

(A) good thermal conductivity
(B) chemically stable
(C) structurally stable
(D) oxygen's low neutron capture cross section

NUCLEAR FUEL MANAGEMENT–7

Individualized fuel pellets are "cupped" as shown.

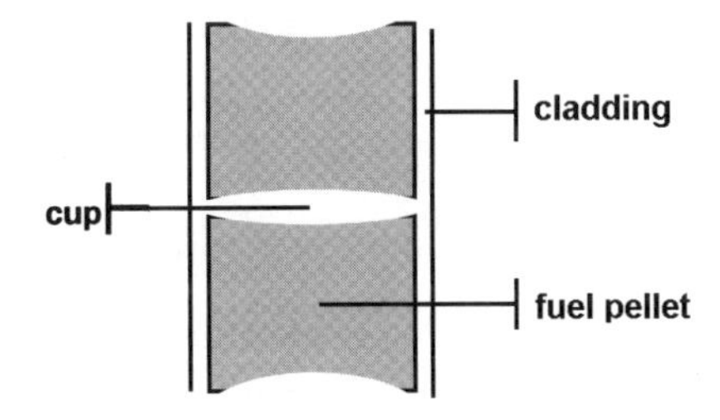

The purpose of the cup is to

(A) allow space for fission products
(B) increase mechanical compression on the outer portion of the pellet to prevent movement
(C) allow for ease of handling during subsequent refueling
(D) trap hydrogen away from Zircaloy cladding to prevent formation of ZrH_2

NUCLEAR FUEL MANAGEMENT–8

During the fabrication of nuclear fuel a small loss, usually less than 1%, occurs. As a result of this loss, more uranium is required with a commensurate increase in the cost at each stage—mining, conversion, enrichment, and fabrication. Representative prices and losses follow.

P_u = price of natural uranium = $60/kgU
P_c = price of conversion = $10/kgU
P_{SWU} = price of separative work unit = $100/SWU
P_f = price of fabrication[10] =$250/kgU
l_c = conversion material loss = 0.2%
l_f = fabrication material loss = 0.5%

Enrichment will be of natural uranium to a 3.3 w/o (i.e., weight percent) U-235 with a tails assay of 0.3%.[11] What is most nearly the total cost per kg-uranium of the fabricated fuel (P_f)?

(A) $800/kgU
(B) $950/kgU
(C) $1050/kgU
(D) $1150/kgU

[10] The cost of fabrication includes the cost of transportation to the site.
[11] The assumption is that the feed uranium is natural uranium with a weight percent of 0.711. This does not have to be the case. Previously enriched uranium which has been used as fuel and then reprocessed will have a higher input percentage depending on the burnup, on the order of 1 w/o.

NUCLEAR FUEL MANAGEMENT–9

A 500 MW_e reactor was operated during a given year as shown.

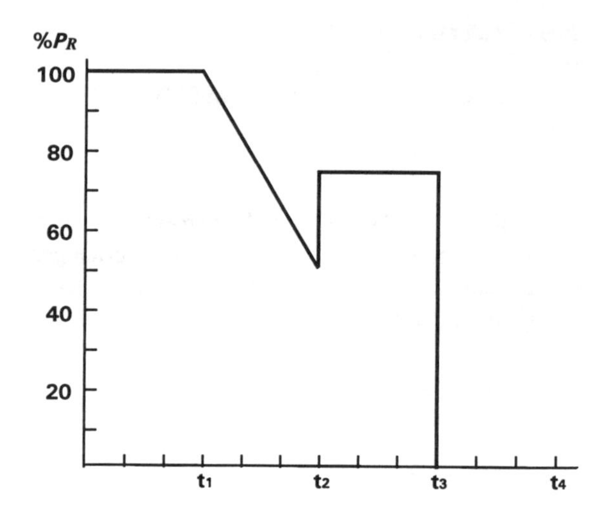

The rated power P_R is given as a percentage of full power. The year is broken into quarter periods t_1, t_2, t_3, and t_4.

What is the capacity factor in percent?

(A) 62.5%
(B) 72.5%
(C) 82.5%
(D) 92.5%

NUCLEAR FUEL MANAGEMENT–10

A 1000 MW_e plant with an efficiency of 30% is initially loaded with 90 000 kg of uranium dioxide (UO_2) fuel. The core operates at full power for one year and then shuts down for refueling. During refueling one-third of the fuel is replaced. What is the average burnup in units of megawatt days per metric ton (MWD/MTU) for the third of the core initially removed?[12]

(A) 4.6×10^3 MWD/MTU
(B) 13.8×10^3 MWD/MTU
(C) 15.3×10^3 MWD/MTU
(D) 45.6×10^3 MWD/MTU

[12] This is a metric ton (i.e., 1000 kg of uranium) represented as "tonne" in some texts and MTU in others, as well as other variations. The reason for the unit change will be explained in the next footnote.

Nuclear Fuel Management (NFM) problems 11–13 are based on the following information and illustration.

A 500 MW_e plant operates with the following power history for a period of one year. Three month periods are shown on the x-axis as t_1, t_2, and so on.

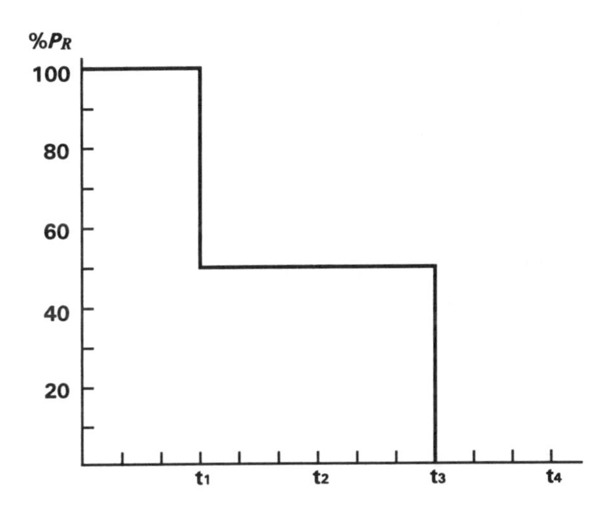

In the three month period from t_3 to t_4 the plant was shutdown for scheduled maintenance and upgrades.

NUCLEAR FUEL MANAGEMENT–11

What is the capacity factor?

(A) 0.25
(B) 0.50
(C) 0.67
(D) 0.75

NUCLEAR FUEL MANAGEMENT–12

What is the availability factor for the time period shown?

(A) 0.25
(B) 0.50
(C) 0.75
(D) 1.00

NUCLEAR FUEL MANAGEMENT–13

What is most nearly the number of effective full power days (EFPD) the plant experienced during this time period?

(A) 90 EFPD
(B) 180 EFPD
(C) 270 EFPD
(D) 360 EFPD

NUCLEAR FUEL MANAGEMENT–14

A certain PWR is a three-batch core. The *excess reactivity* in the core $\rho_N(t)$ is governed by the following expression[13]

$$\rho_N(t) = \sum_{n-1}^{N} \frac{\left(1 - \mathrm{BU}_N^n(t)\right)\left(\rho_1\right)}{N}$$

N = number of batches in the core
n = cycle number
ρ_1 = excess reactivity in a one batch core

The final burnup is governed by

$$\mathrm{BU}_N(T) = \left(\frac{2N}{N+1}\right)\left(\mathrm{BU}_1(t)\right)$$

T = time period of concern
BU_N = the final burnup of a batch in a N batch core, once that batch has been in the core for N cycles[14]

For the same initial enrichment, increasing the number of batches in a core does which of the following:

(A) decrease the time between refuelings
(B) increase the availability factor
(C) increase the final burnup by one and a half times over a single batch core
(D) A and C above

[13] The equation assumes a linear relationship between reactivity and burnup. Additionally, the assumption is made that the total core reactivity, at any time in a cycle, is equal to the average of the sums of the individual reactivities.
[14] The units are MWD/MTU or similar units.

Nuclear Fuel Management (NFM) problems 15 and 16 are based on the following information and illustration.

A 1000 MW_e PWR operates at full power and 25% efficiency for 11 months. The decay heat removal (DHR) requirements must be determined to ensure adequate cooling. One method of decay heat generation determination is to utilize the following formula (or one similar).[15]

$$\frac{P(t_o, t_s)}{P_o} = \frac{P(t_s)}{P_o} - \frac{P(t_o + t_s)}{P_o}$$

P = power generated
P_o = thermal power
t_o = operating time in seconds
t_s = time since shutdown in seconds

As decay heat generation is directly related to the number of fissions and thus the thermal power, a normalized graph, this one adapted from the standard ANS-5, can be generated as shown.

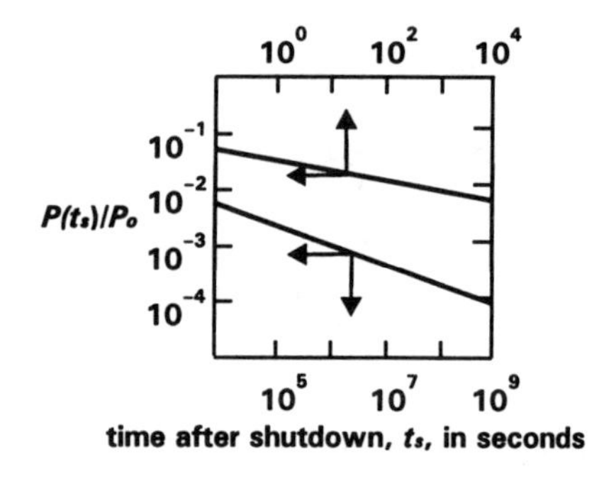

[15] Several formulas exist. Some use coefficients for each term whose value depends on plant conditions and operating history. The formula used depends on regulatory requirements and the accuracy desired.

NUCLEAR FUEL MANAGEMENT–15

What is the decay heat generation power removal requirement 24 hours after shutdown?

(A) 4 MW
(B) 17 MW
(C) 70 MW
(D) 280 MW

NUCLEAR FUEL MANAGEMENT–16

By approximately what factor is the decay heat generation requirement greater immediately upon shutdown as compared to the 24 hour period calculated in the Prob. 15?

(A) 2
(B) 10
(C) 15
(D) 100

Nuclear Fuel Management (NFM) problems 17–20 are based on the following information and tables and represent a partial excerpt from 10CFR61.[16]

activity concentrations used to classify low-level wastes (LLW)

Table A

nuclide	concentration (Ci/m^3)
Ni-59	220
Tc-99	3
I-129	0.08
alpha-emitting transuranic waste (TRU) with a $T_{1/2}$ >5 yr	100 nCi/g
Pu-241	3500 nCi/g

Table B

	concentration (Ci/m^3)		
nuclide	column 1	column 2	column 3
$T_{1/2}$ < 5 yr	700	*	*
^3H	40	*	*
Ni-63	3.5	70	700
Sr-90	0.04	150	7000
Cs-137	1	44	4600

*No specific limit is established. Heat generation or external radiation may limit their concentration. These wastes are class B unless other considerations make them class C.

Classification guidance using the above tables can be found in Table 2.3 of App. 2.B.

NUCLEAR FUEL MANAGEMENT–17

A LLW package is determined to contain Ni-59 at $0.02\ mCi/cm^3$ and Ni-63 at $1\ Ci/cm^3$. What is the classification of this waste?

(A) class A
(B) class B
(C) class C
(D) GTCC

[16] Title 10 of the Code of Federal Regulations, Chapter 61. Consult the latest revision to determine up-to-date requirements.

NUCLEAR FUEL MANAGEMENT–18

A LLW package contains Sr-90 in a concentration of $75\ Ci/m^3$ and Cs-137 in a concentration of $25\ \mu Ci/cm^3$. What is the classification of this package?

(A) class A
(B) class B
(C) class C
(D) GTCC

NUCLEAR FUEL MANAGEMENT–19

A LLW package contains Tc-99 at a concentration of $0.02\ Ci/m^3$ and tritium at a concentration of $45\ Ci/m^3$. What is the classification of this waste?

(A) class A
(B) class B
(C) class C
(D) GTCC

NUCLEAR FUEL MANAGEMENT–20

A LLW package with dimensions of 12 in × 12 in × 8 in contains 20 000 μCi of ^{32}P. What is the classification of this package?

(A) class A
(B) class B
(C) class C
(D) GTCC

Nuclear Fuel Management (NFM) problems 21 and 22 are based on the following information and equation.

There are five general categories of radioactive wastes:

High-Level Wastes (HLW). Defined in the Nuclear Waste Policy Act as the "highly radioactive material from reprocessing spent fuel...."[17]
Transuranic Wastes (TRU). Defined as alpha emitting isotopes with an atomic number greater than 92, half-lives greater than five years and activity concentrations greater than 100 nCi/g of waste.[18]
Low-Level Waste (LLW). Defined as material that is not HLW, spent nuclear fuel, TRU, or by-product material.[19]
Uranium mill tailings. These are treated differently from other radioactive wastes since they are not transported from their generation point.
Naturally Occurring and Accelerator produced Radioactive Materials (NARM). These wastes are not regulated by the Nuclear Regulatory Commission (NRC) but by individual states and generally classified as LLW.
Radioactive wastes generate heat. Spent fuel is often stored in a pool upon initial removal from a reactor in order to remove this heat.[20] One method of determining decay heat generation is using the equation found in an American Nuclear Society standard 5.1 for decay heat generation calculation. Specifically,

$$\frac{P(t,\infty)}{P_o} = \mathrm{A}t^{-\mathrm{a}}$$

$P(t,\infty)$ = decay power at t seconds after shutdown, following operation at P_o for an infinite time
t = time in seconds following reactor shutdown
A,a = constants

constants "A" and "a" of ANS 5.1 Standard

time interval (sec)	A	a
$0.1 < t < 10$	0.0603	0.0639
$10 < t < 150$	0.0766	0.181
$150 < t < 4 \times 10^6$	0.130	0.283
$4 \times 10^6 < t < 2 \times 10^8$	0.266	0.335

NUCLEAR FUEL MANAGEMENT–21

An 1150 $\mathrm{MW_e}$ PWR with a plant efficiency of 30% has been shutdown for 30 days after operating for an extended time at rated power. If the spent fuel is to be moved to a pool, what is the heat removal capacity required?

(A) 1.2×10^2 Btu/hr
(B) 2.1×10^2 Btu/hr
(C) 7.6×10^6 Btu/hr
(D) 2.6×10^7 Btu/hr

NUCLEAR FUEL MANAGEMENT–22

Approximately 4 mCi of Pu-240 is to be disposed of in a package containing 1.1 kg of material with essentially negligible activity. What is the classification of this waste?

(A) LLW class A
(B) LLW class C
(C) TRU
(D) either B or C

[17] There are three general groups of radioisotopes in spent fuel: fission products, *actinides*, and activation products.
[18] This is actually a subset of HLW. Further, TRUs have been defined as a category only in the U.S.
[19] Questions in the Nuclear Fuel Management section more clearly define low level wastes. See Table 2.3 in App. 2.B for additional information.
[20] Dry cask storage is also used but only after the decay heat generation is reduced to a rate where the heat removal capability of the cask is capable of maintaining the required temperature.

NUCLEAR FUEL MANAGEMENT–23

Current plans call for the disposal of HLW in a deep geological repository. This disposal method will also be used for TRU, for which there is no essential difference with HLW regarding treatment and disposal. The largest portion of the HLW and TRU is in liquid form. Disposal is expected to occur in solid form. Possible methods follow.

α	calcination
β	cementation
γ	vitrification

Partial descriptions of the methods follow.

1	mix with cement; pour into a container and allow to dry
2	mix with glass frit
3	heat until completely dry

Match the methods given with their descriptions. Also, specify the preferred method.

(A) α1, β2, γ3/calcination
(B) α2, β1, γ3/vitrification
(C) α2, β3, γ1/cementation
(D) α3, β1, γ2/vitrification

Nuclear Fuel Management (NFM) problems 24 and 25 are based on the following information and equation.

Transportation of radioactive materials is governed by the U.S. Department of Transportation (DOT);[21] the Nuclear Regulatory Commission (NRC);[22] and the U.S. Postal Service.[23] International transportation is further governed by International Atomic Energy Agency (IAEA) regulations.

Differentiation between HLW and LLW does not occur. Instead, packaging requirements depend on the isotopes in a given package and the activity associated with those isotopes. Numerous definitions of material and classifications exist.

One such classification is low specific activity (LSA) material. A partial definition of LSA material follows.

LSA = material which has essentially uniformly distributed radioactivity in which the average concentration per gram does not exceed the requirements of 10CFR71

A partial table from 10CFR71 for selective isotopes follows.

isotope	A1 special form (Ci)	A2 normal form (Ci)
Co-60	7	7
Cs-137	30	10
Pb-210	100	0.2
Ra-226	10	0.05
U-233	100	0.1

A2 = the maximum activity of radioactive material, other than special form radioactive material, permitted in a type A package

A material is considered LSA material if it meets the following requirements from 10CFR71 regarding the average concentration per gram of activity:

(i) no more than 0.0001 mCi from radionuclides with A2 activity ≤ 0.05 Ci

[21] See 49CFR100–199.
[22] See 10CFR71.
[23] See 39CFR124.

(ii) no more than 0.005 mCi from radionuclides with A2 activity more than 0.05 Ci but less than 1 Ci
(iii) no more than 0.3 mCi from radionuclides with A2 activity more than 1 Ci

If a mixture of radionuclides is to be classified as LSA, it must meet the following condition (know as the ratio rule).

$$\frac{A}{0.0001}+\frac{B}{0.005}+\frac{C}{0.3}<1.0$$

A = the total activity in mCi/g of all nuclides with an A2 activity $\leq$ 0.05 Ci
B = the total activity in mCi/g of all nuclides with an A2 activity > 0.05 Ci but < 1.0 Ci
C = the total activity in mCi/g of all nuclides with an A2 activity $\geq$ 1.0 Ci

Consider the following radioactive material for transport.

	material	estimated concentration per gram
I	Ra-226	0.1 μCi
II	Pb-210	50 μCi
III	U-233	500 pCi
IV	Cs-137	3500 μCi

NUCLEAR FUEL MANAGEMENT–24

If the material is shipped separately, which packages can be shipped as LSA material in a type A package?

(A) I and II
(B) I and III
(C) II and IV
(D) I, II, and III

NUCLEAR FUEL MANAGEMENT–25

Which combination of materials can be transported together and retain the LSA designation?

(A) no combination
(B) I and II
(C) I and III
(D) II and III

3 NUCLEAR RADIATION

Nuclear Radiation (NR) problems 1–3 are based on the following information and illustration. [1]

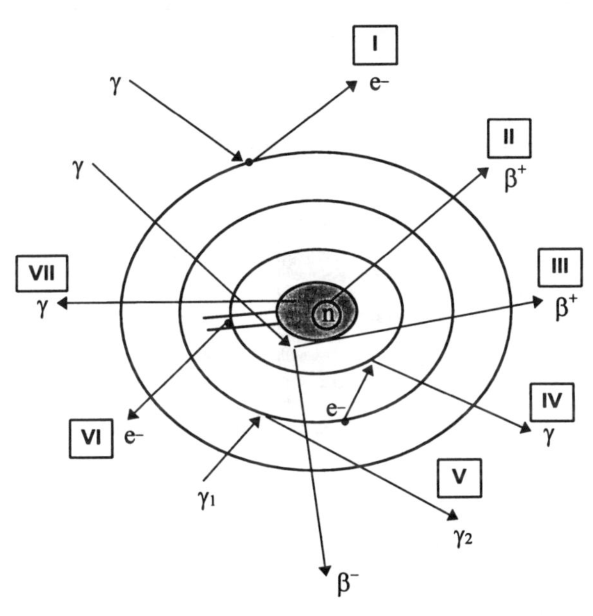

γ = photons [subscripts indicate different frequencies]
e– = electrons
β = beta particles [electrons/positrons]

[1] The orbits of the electrons are shown as circular. This is certainly not strictly true. Indeed, one should speak of the orbits in terms of position probability. While not required for the nuclear professional engineer examination, an understanding of the atoms at this level means having an understanding of quantum theory. See the Reference section for more information.

NUCLEAR RADIATION–1

Which process is referred to as the photoelectric effect?

(A) I
(B) II
(C) IV
(D) VI

NUCLEAR RADIATION–2

Which process is referred to as Compton scattering?

(A) I
(B) IV
(C) V
(D) VI

NUCLEAR RADIATION–3

Which process requires a minimum of 1.02 MeV?

(A) II
(B) III
(C) VI
(D) VII

Nuclear Radiation (NR) problems 4 and 5 are based on the following information.

The total cross section for γ-ray interaction for the three most important effects with regard to shielding is

$$\sigma = \sigma_{pe} + \sigma_{pp} + \sigma_c$$

σ_{pe} = microscopic cross section for photoelectric effect
σ_{pp} = microscopic cross section for pair production
σ_c = microscopic cross section for Compton scattering[2]

Multiplying the cross section by the atom density N would normally give the macroscopic cross section Σ as the result.

By tradition, the macroscopic cross sections of gamma ray interaction are called *attenuation coefficients* and symbolized by μ.[3] Therefore,

$$\mu = N\sigma = \mu_{pe} + \mu_{pp} + \mu_c$$

When divided by the density of the material these coefficients become the *mass attenuation coefficients* and are routinely tabulated as such.[4]

$$\frac{\mu}{\rho} = \frac{\mu_{pe}}{\rho} + \frac{\mu_{pp}}{\rho} + \frac{\mu_c}{\rho}$$

To account for the energy deposited, which is closely related to the biological damage, the average energy of the recoiling electron in Compton scattering is accounted for and a *mass absorption coefficient* is determined.[5]

$$\frac{\mu_a}{\rho} = \frac{\mu_{pe}}{\rho} + \frac{\mu_{pp}}{\rho} + \frac{\mu_{c,a}}{\rho}$$

The mass absorption coefficient is used to determine radiation intensity I and an energy deposition rate E_{dr}.

NUCLEAR RADIATION–4

Lead with a mass thickness of 86.49 g/cm² is to be used to shield against 1.33 MeV γ-rays of intensity 3×10^5 s^{-1}.[6]

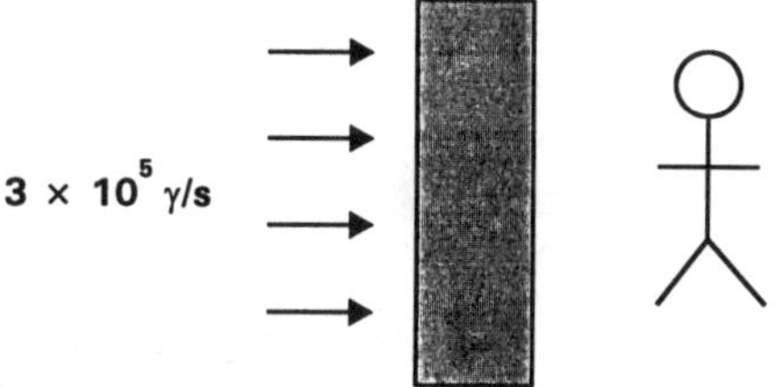

What is the approximate intensity of the gamma rays near the worker which have not interacted with the lead?[7]

(A) 2.6×10^2 s^{-1}
(B) 2.2×10^3 s^{-1}
(C) 3.0×10^5 s^{-1}
(D) 4.0×10^7 s^{-1}

[2] In the Compton scattering interaction the photon (or γ-ray) interacts with individual electrons. Therefore, a cross section per electron σ_{ce} exists. The total probability for any given atom is the number of electrons in the atom Z_e multiplied by the individual probabilities σ_{ce}. In other words, $\sigma_c = Z_e\sigma_{ce}$.

[3] This represents the probability per unit path length of an interaction (i.e., photoelectric effect (PEE); pair production (PP) and Compton scattering (CS)). Physicists call these items linear absorption coefficients. Not all the gamma rays are absorbed however, and this must be taken into account in shielding problems by using buildup factors.

[4] These are useful in that the ratio μ/ρ is essentially constant over the range of energies where Compton scattering dominates (i.e., from approximately 0.5 MeV to 5 MeV) and thus the effectiveness of a shield varies only with its mass thickness (i.e., ρx). See the Reference section for more information.

[5] These are called energy absorption coefficients in some books and given the symbol μ_{en}.

[6] The γ-ray from cobalt 60 decay has an energy of approximately 1.33 MeV.

[7] The gamma rays that have not interacted are not the only photons present on the worker's side. Compton scattering results in scattered photons, *x*-rays may follow the photoelectric effect, and annihilation radiation follows pair production, all which may be present depending upon where they were generated in the lead. These effects can be factored into the appropriate equations using buildup factors as will be seen in follow-on problems. This question is a rough calculation since photons that have interacted have been ignored; thus, the need to use a buildup factor has been eliminated.

NUCLEAR RADIATION–5

A 15 kg carbon steel drum comprised of 95 w/o (i.e., weight percent) iron and 5 w/o silicon is used to store a radioactive liquid.[8] The activity of the liquid on the interior surface of the drum is estimated to be approximately 2×10^{11} $\gamma/\text{cm}^2\cdot\text{s}$ with an average gamma ray energy of 1.0 MeV.

What is the energy deposition rate into the drum?

(A) 1–5 W
(B) 5–10 W
(C) 10–15 W
(D) 15–20 W

Nuclear Radiation (NR) problems 6 and 7 are based on the following information.

selected properties of cobalt	
density Co-59	8.8 g/cm3
σ_a	37.2 barns
AW	58.9332
Co-60 half-life	5.271 yr

NUCLEAR RADIATION–6

A small cylindrical support pin with a radius of 0.5 cm and length of 1.0 cm is made from Co-60. The pin is exposed to a monoenergetic neutron flux.

$$\phi = 8 \times 10^5 \text{ neutrons/cm}^2 \cdot \text{s}$$

In one year of steady state operation, what is the number of atoms activated in this pin?

(A) 2.7×10^6 atoms
(B) 6.6×10^{13} atoms
(C) 9.0×10^{13} atoms
(D) 12.9×10^{20} atoms

NUCLEAR RADIATION–7

What is the approximate activity of the pin in the Prob. 6 at the one-year point, in becquerels?

(A) 2.76×10^5 Bq
(B) 3.55×10^5 Bq
(C) 8.68×10^{12} Bq
(D) 1.18×10^{13} Bq

NUCLEAR RADIATION–8

A sample of primary coolant is drawn from a pressurized water reactor (PWR). It is depressurized and the gaseous constituents contained in an enclosed chamber with a 2 mm steel cover on one side that allows for activity measurements. The initial activity measurement A_0 is 111 counts per second (cps) above normal.

In discussions the following potential causes are surmised.

(A) The activity is a result of manganese and iron from corrosion products due to the shock from a recent startup following an extended refueling.

(B) The activity is due to naturally occurring nitrogen from the following reaction.[9]

$$^{16}\text{O}[\text{n}, \text{p}]^{16}\text{N}$$

(C) The activity may be from xenon 135 indicating a potential fuel element failure may have occurred while shutdown.

(D) During the recent pressurization following refueling, air may have been added to the plant resulting in argon 41 activity.

Following the discussion, four hours after the initial sample, a second activity is taken on the original sample and a γ-ray spectrometer measurement is taken.

$$A_4 = 22 \text{ cps}$$

$$\text{detected gamma} = 1.3 \text{ MeV } \gamma$$

[8] Silicon makes the steel harder. Its main purpose here, however, is as the impetus to explain the use of the mass absorption coefficient in mixtures of material.

[9] The symbology shown indicates an oxygen 16 atom absorbing a neutron and then decaying by expulsion of a proton to become nitrogen 16.

What is the source of the activity?

(A) Mn-56 and Fe-59
(B) N-16
(C) Xe-135
(D) Ar-41

Nuclear Radiation (NR) problems 9–12 are based on the following information which represent a portion of a chart of the nuclides.[10]

Z ↑			
	Ba-136	Ba-137	Ba-138
	Cs-135	Cs-136	Cs-137
	Xe-134	Xe-135	Xe-136

N ⟶

N = number of neutrons
Z = number of protons

Now consider the following portion of the chart of the nuclides. The shaded block represent the original nucleus. The other blocks are arbitrarily labeled to represent the indicated position on the chart.

			IX	XIII
	III	VII	X	XIV
	IV	(shaded)	XI	
I	V	VIII	XII	
II	VI			

NUCLEAR RADIATION–9

If the original nucleus decays by the emission of an alpha particle, what position will it then occupy on the chart?

(A) II
(B) V
(C) X
(D) XIII

[10] An actual chart contains significantly more information. It is recommended one become familiar with such a chart prior to the nuclear profession engineer examination. See the Reference section for more information.

NUCLEAR RADIATION–10

If the original nucleus decays by β^- emission, what position will it then occupy on the chart?

(A) III
(B) IV
(C) VII
(D) XI

NUCLEAR RADIATION–11

Consider the following portion of a chart of the nuclides.

Z ↑				
	U-232	U-233	U-234	U-235
	Pa-231	Pa-232	Pa-233	Pa-234
	Th-230	Th-231	Th-232	Th-233
	Ac-229	Ac-230	Ac-231	Ac-232

N ⟶

The original nucleus is U-235. The U-235 undergoes alpha decay; the resulting nucleus β^- decays. This nucleus then absorbs two neutrons. What is the resultant element?

(A) Ac-231
(B) Th-232
(C) Pa-233
(D) U-234

NUCLEAR RADIATION–12

Consider the symbology introduced earlier.

$$X[\text{particle absorbed; particle emitted}]Y$$

X = original nucleus
Y = resulting nucleus

If the original nucleus undergoes a neutron capture reaction [n,γ], what is its final position on the chart?

(A) I
(B) IV
(C) XI
(D) XII

NUCLEAR RADIATION–13

Consider the following graph and table.[11]

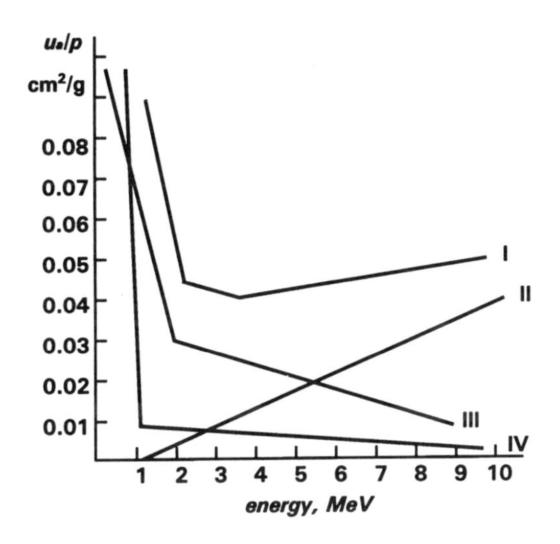

Match each line (i.e., I, II, III, IV) with the associated effect given below.

designation	effect
A	photoelectric effect
B	Compton scattering
C	pair production
D	total

(A) I A II B III C IV D
(B) I A II C III B IV D
(C) I D II B III A IV C
(D) I D II C III B IV A

[11] The graph is shown with linear approximations for ease of use only. See the Reference section for texts with more detailed information.

Nuclear Radiation (NR) problems 14 and 15 are based on the following information.

traditional unit	symbol	unit of	SI unit	notes
roentgen[12]	R[a]	exposure	C/kg	1 R = 2.58×10^{-4} C/kg
rad	rad	absorbed dose	Gy	1 gray = 1 J/kg = 100 rad
rem	rem	absorbed dose equivalent	Sv	1 sievert = 100 rem

[a] The symbol R used for roentgen is often also used to indicate rem. Caution is advised. In the SI system, this problem does not exist.

quality factors	
x-ray; γ-ray; *β*-ray	1
thermal neutrons	2
neutrons with unknown energy; protons	10
alpha particles	20

NUCLEAR RADIATION–14

A 10 mR/hr 1.0 MeV gamma radiation field exists in a localized area of a nuclear power plant. At approximately what rate are ions being produced in the air in this area?

(A) 3×10^{-7} $cm^{-3}\cdot s^{-1}$
(B) 30×10^{-7} $cm^{-3}\cdot s^{-1}$
(C) 60 $cm^{-3}\cdot s^{-1}$
(D) 6000 $cm^{-3}\cdot s^{-1}$

NUCLEAR RADIATION–15

What is the dose equivalent for a person who remains in the gamma ray field described in the Prob. 14 for 8 hours?

(A) 80 mGy
(B) 800 μGy
(C) 80 mSv
(D) 800 μSv

[12] Additional useful information follows. One *esu* = 3.33×10^{-10} C. The mass of one cm^3 of dry air at STP = 0.001293 g.

Nuclear Radiation (NR) problems 16–18 are based on the following information.

Consider the following potential shielding materials.

designation	material
I	iron
II	lead
III	water
IV	paraffin
V	polyethylene

Consider the following list of shield material advantages.

designation	advantage
a	inexpensive
b	high density
c	high strength
d	no neutron activation problems

Consider the following list of shield material disadvantages.

designation	disadvantage
i	activates in neutron fluxes
ii	toxic
iii	combustible; fuel for a fire
iv	potential loss of shield due to leakage

NUCLEAR RADIATION–16

What are the advantages of lead (II) and polyethylene (V) for use as shielding?

(A) II a V a
(B) II b V d
(C) II c V a
(D) II d V d

NUCLEAR RADIATION–17

What are the disadvantages of iron (I) and paraffin (IV)?

(A) I i IV iii
(B) I ii IV iii
(C) I iii IV iv
(D) I iv IV iv

NUCLEAR RADIATION–18

What is a tenth-value layer for the uncollided flux in ordinary concrete for a 1.0 MeV gamma ray beam?[13] The density of ordinary concrete is approximately 2.3 g/cm^3.

(A) 5 cm
(B) 10 cm
(C) 15 cm
(D) 20 cm

NUCLEAR RADIATION–19

An isotropic point source is used as a calibration test source for radiation detection instruments. It is stored in a lead container as shown.

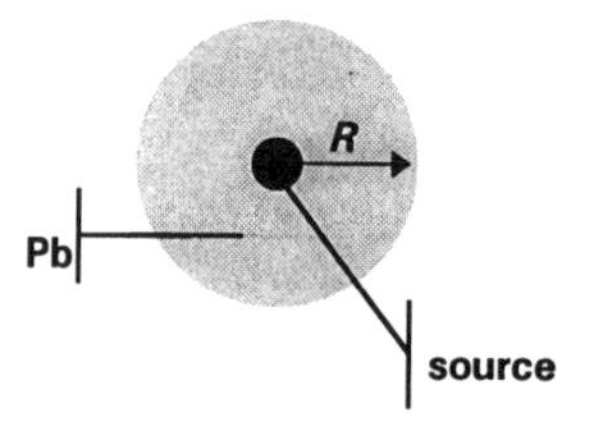

$R = 1$ cm
$\rho_{Rb} = 11.35$ g/cm^3
$E_\gamma = 1.0$ MeV

A quick calculation assuming spherical spreading of the gamma flux indicates the exposure rate to be 10 mR/hr without the shield present.[14]

Using the Taylor form of the point isotropic buildup factor, what is the expected exposure rate at the surface of the lead shield?

(A) 2 mR/hr
(B) 5 mR/hr
(C) 6 mR/hr
(D) 10 mR/hr

[13] The tenth-value layer is also called the tenth-thickness. Half-value thicknesses are commonly tabulated as well. The reduction of a given beam of radiation to five percent of its original value is also a common reference.
[14] An operational way of determining this exposure would be to simply measure it at the desired distance. One would then calculate the thickness of the shielding required for a desired exposure rate. A programmable calculator or semi-log paper is helpful. See the Reference section for texts which provide further examples of calculations of this sort.

NUCLEAR RADIATION–20

A radioactive spill occurs on the shielded floor of a sampling laboratory. The situation is as shown.

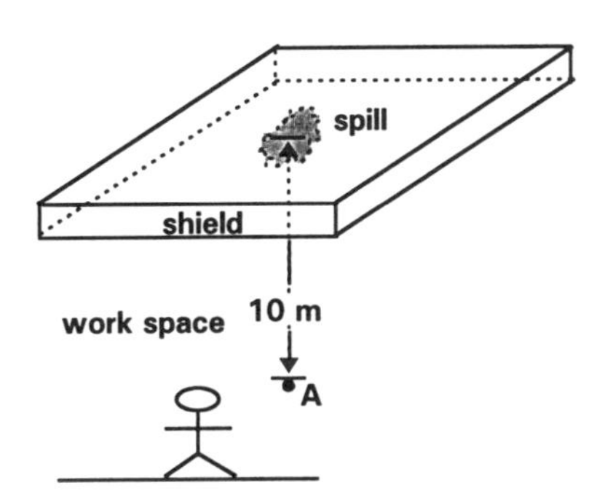

Assuming the spill is approximately circular, the solution for the absorbed dose equivalent at point A is of which form?

(A) 1/(10 m)
(B) Sievert integral[15]
(C) $1/(10\text{ m})^2$
(D) E_1 [exponential integral]

NUCLEAR RADIATION–21

One millicurie of naturally occurring Ra-226 finds its way into a 5 000 000 gallon community water supply tank. does the resulting water violate EPA standards for drinking water and what is the primary hazard from this element?[16]

(A) yes; internal α
(B) yes; external γ
(C) no; internal α
(D) no; external γ

[15] This is also called the secant integral.
[16] This element is used in industrial radiography and for medical purposes.

NUCLEAR RADIATION–22

Reactor siting is an important aspect of the design and licensing of a nuclear plant. What factors are considered by the NRC in evaluating a site?[17]

(A) reactor design and population density
(B) site physical characteristics
(C) engineering safeguards
(D) all of the above

NUCLEAR RADIATION–23

Numerous surface decontamination methods exist. Consider the following.

designation	method/material
I	vacuum cleaning
II	water
III	steam
IV	abrasion

Each method has advantages and disadvantages and must be picked based on the nature of the problem to be resolved. Consider the following.

designation	remarks
a	Activity is reduced to as low as desired. Contamination can spread.
b	Activity is reduced by approximately 90%, and this method can be used on oily surfaces.
c	This method requires the use of a filter.
d	Activity is reduced by approximately 50%.

Match each of the methods with the appropriate remarks.

(A) I a II b III c IV d
(B) I c II d III a IV b
(C) I a II b III d IV c
(D) I c II d III b IV a

[17] See 10CFR100 (i.e., title 10, Code of Federal Regulations, chapter 100) for additional information. The Atomic Energy Act of 1954, court decisions and Regulatory Guides also provide information on reactor siting.

4 NUCLEAR THEORY

NUCLEAR THEORY–1

Approximately how much energy is released in the conversion/annihilation of 1 lb of mass?

(A) 400 kw/hr
(B) 4000 kw/hr
(C) 11 000 000 kw/hr
(D) 11 000 000 000 kw/hr

NUCLEAR THEORY–2

What is the approximate binding energy per nucleon of U-235?

(A) 1.5 MeV
(B) 7.8 MeV
(C) 235 MeV
(D) 1700 MeV

NUCLEAR THEORY–3

The energy distribution among gas atoms or molecules tends to follow the Maxwellian distribution given by

$$N(E) = \left(\frac{2\pi N}{(\pi kT)^{3/2}}\right)\left(E^{1/2}\right)\left(e^{-\frac{E}{kT}}\right)$$

$N(E)$ = number of particles with energy E
k = Boltzmann's constant
T = temperature in kelvin

A plot of the energies takes the following shape.

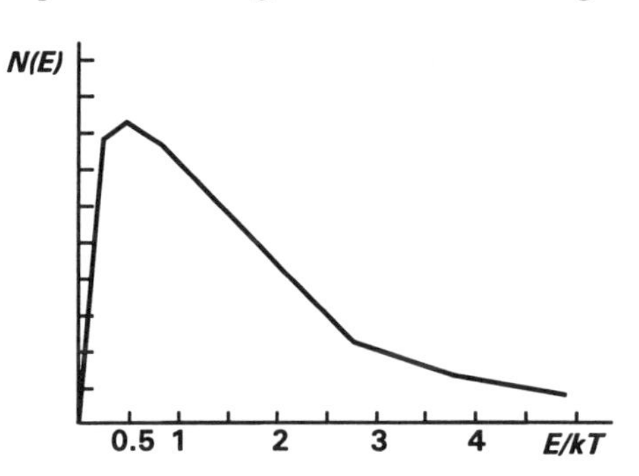

Above room temperature (i.e., approximately 300 K) the equation can be used for liquids and solids with minimal error. The most probable energy is

$$E_p = \frac{1}{2}kT$$

The average energy is

$$\overline{E} = \frac{3}{2}kT$$

Assuming a reference temperature T_0 of 293.61 K (20°C), what is the energy associated with the term kT?[1] What is the speed of a neutron whose energy is given by the term kT?

(A) 0.0253 eV; 2200 m/s
(B) 0.0253 eV; 1560 m/s
(C) 0.511 MeV; 2200 m/s
(D) 1.02 MeV; 1560 m/s

[1] The value of T_0 (20°C) is the temperature to which many nuclear parameters are referenced.

NUCLEAR THEORY–4

A point source emits 10^{13} neutrons/s into a water moderator as shown.

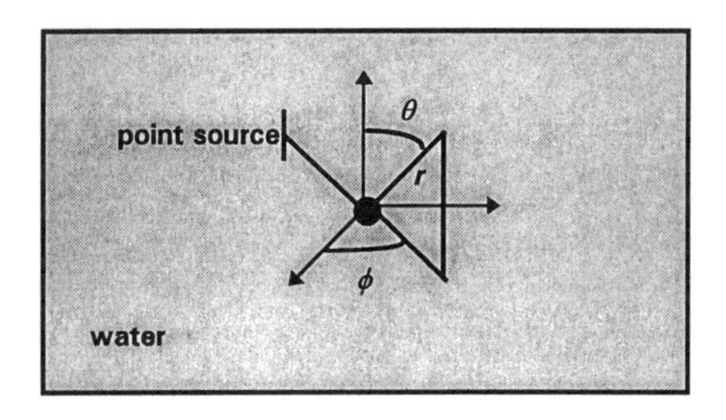

The neutron flux is given by[2]

$$\phi(r) = \frac{Se^{-\frac{r}{L}}}{4\pi Dr}$$

S = source strength neutrons/s
L = diffusion length
D = diffusion coefficient
r = radial position

For water, approximate values for the defined quantities are

$$L = 3 \text{ cm}$$
$$D = 0.3 \text{ cm}$$

What is the net rate of neutrons flowing out of a sphere of radius 2 m surrounding the point source? Use Fick's law (i.e., the diffusion approximation).

(A) 8×10^{-15} s-1
(B) 9×10^{-12} s-1
(C) 9×10^{12} s-1
(D) 8×10^{15} s-1

[2] This formula assumes an infinite moderator and uses as its basis neutron diffusion (i.e., Fick's law). See the Reference section for more information.

NUCLEAR THEORY–5

The following equation represents the general form of the neutron continuity equation.

$$\frac{\delta n}{\delta t} = S - \Sigma_a \phi - \text{div}\vec{J}$$

Which term represents leakage of neutrons from the volume under consideration?

(A) S
(B) $\Sigma_a\phi$
(C) div$\boldsymbol{J}$
(D) B or C

NUCLEAR THEORY–6

The diffusion equation previously used assumed monoenergetic neutrons (i.e., it can be considered a one-group equation). For thermal reactors at least two groups of neutron energies should be used, fast and thermal.[3]

The energy groups are organized as follows.

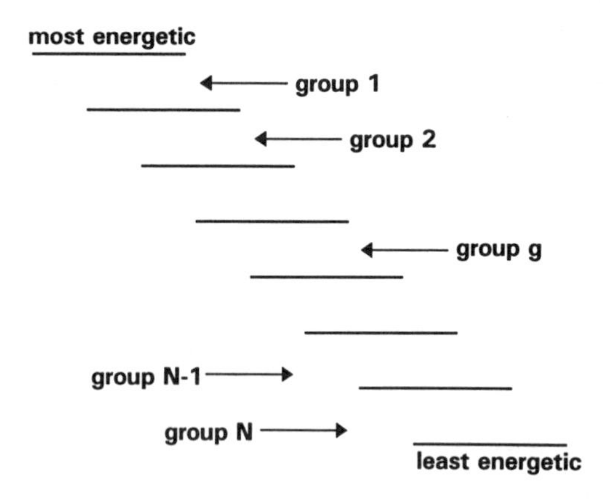

[3] These calculations can involve many neutron energy groups and are termed multi-group calculations which normally involve the use of computers to aid in determination of the fluxes. This is a much more exact method of determining reactor properties than using one- and two-group diffusion results.

Consider a certain fast reactor, fueled with U-238 and moderated by Na-23. The neutrons are separated into six groups with the corresponding cross sections shown.[4]

group	σ_{tr} barns	σ_r barns	$\sigma_{g\to g+1}$ barns	ϕ_g cm-3·s-1
1	6.3	0.04	2.44	1×10^{13}
2	7.5	0.07	1.02	3.5×10^{13}
3	0.11	0.10	0.95	4×10^{14}
4	0.11	0.15	0.63	7×10^{14}
5	0.14	0.23	0.40	1×10^{15}
6	0.16	0.61	–	5×10^{16}

Sodium density is 0.97 g/cm^3 for Na-11.[5]

What is most nearly the rate at which neutrons are scattered from the second into the third group?

(A) 6×10^{11} cm-3·s-1
(B) 9×10^{11} cm-3·s-1
(C) 10×10^{12} cm-3·s-1
(D) 9×10^{13} cm-3·s-1

NUCLEAR THEORY–7

What is the thermal flux corresponding to a neutron density of 10^{13} neutrons/cm^3 at 293.61 K?

(A) 2.2×10^{13} cm-3·s-1
(B) 2.5×10^{15} cm-3·s-1
(C) 2.2×10^{18} cm-3·s-1
(D) 2.5×10^{18} cm-3·s-1

NUCLEAR THEORY–8

What is the ratio of the 2200 m/s flux to the thermal flux given an operating temperature of 597 K?

(A) 0.5
(B) 0.6
(C) 0.7
(D) 0.8

[4] The values given are approximate. Additionally, only g→g+1 cross sections are shown. Cross sections for g→g+2, g→g+3, and so on could be used. Scattering from a lower energy group into a higher energy group (e.g., g+3→g+2) must also be accounted for in complete calculations. Final note, the fluxes are arbitrary.
[5] If the term normal density is used it refers to the atom density × 10^{-24} (i.e., atoms/cm3) for the material of interest at standard conditions.

NUCLEAR THEORY–9

The one-group diffusion equation is[6]

$$D\nabla^2\phi - \Sigma_a\phi + S = 0 \text{ or}$$
$$D\nabla^2\phi - \Sigma_a\phi = -S$$

The source term S is due to fissions.[7] Thus,

$$\begin{aligned} S &= \eta\Sigma_{a,\text{fuel}}\phi \\ &= \eta\left(\frac{\Sigma_{a,\text{fuel}}}{\Sigma_a}\right)\Sigma_a\phi \\ &= \eta f\Sigma_a\phi \end{aligned}$$

η = average number of fission neutrons emitted per neutron absorbed in the fuel, called the regeneration factor

$$\eta = \nu\left(\frac{\sigma_f}{\sigma_a}\right)$$

ν = average number of neutrons emitted during fission, both prompt and delayed
$\Sigma_{a,\text{fuel}}$ = macroscopic absorption cross section in the fuel
Σ_a = macroscopic absorption cross section in all material
f = fuel utilization[8]

The infinite multiplication factor is the number of neutrons in one generation divided by the number of neutrons in the preceding generation.

$$k_\infty = \frac{\eta f\Sigma_a\phi}{\Sigma_a\phi} = \eta f$$

[6] The explanation and equations that follow are applicable to one-group (i.e., fast) reactors. The principles can be extended to thermal reactors.
[7] This is true for a critical reactor. Several sources of neutrons exist for a subcritical reactor with a minor number of them coming from spontaneous fission.
[8] This term is called the thermal utilization factor in thermal reactors.

Substituting into the diffusion equations gives

$$D\nabla^2\phi - \Sigma_a\phi = -k_\infty\Sigma_a\phi$$

$$D\nabla^2\phi + (k_\infty - 1)\Sigma_a\phi = 0$$

$$\nabla^2\phi + (k_\infty - 1)\frac{\Sigma_a}{D}\phi = 0$$

$$\nabla^2\phi + \frac{(k_\infty - 1)}{L^2}\phi = 0$$

The term $L^2 = D/\Sigma_a$ and is called the diffusion area.

Define the *buckling* B^2 as

$$B^2 = \frac{k_\infty - 1}{L^2}$$

Substituting gives the final result.

$$\nabla^2\phi + B^2\phi = 0$$

This is called the one-group reactor equation. The buckling is determined by the material properties of the system only.[9]

What is the value for the fuel utilization f and the infinite multiplication factor k_∞ for a homogeneous mixture of U-235 and iron in which the uranium is 1 w/o (i.e., weight percent) of the total?

(A) f 0.394 k_∞ 0.867
(B) f 0.654 k_∞ 1.43
(C) f 0.922 k_∞ 2.03
(D) f 11.9 k_∞ 26.1

Nuclear Theory (NT) problems 10 and 11 are based on the following information and equations.

A spherical reactor shape is often used as a first estimate of expected properties. Recall, the one-group diffusion equation is

$$\nabla^2\phi + B^2\phi = 0$$

Using this equation and solving for the flux in a sphere of radius R gives the following.

buckling	flux	constant A
$\left(\frac{\pi}{R}\right)^2$	$A\left(\frac{1}{r}\right)\sin\left(\frac{\pi r}{R}\right)$	$\frac{P}{4R^2E_R\Sigma_f}$

E_R = recoverable energy per fission in Joules per fission
Σ_f = macroscopic fission cross section
P = operating power (which determines the magnitude of the flux)
$(\pi/R)^2$ = the first eigenvalue squared $(B_1)^2$ which is the solution to the flux equation[10]

Since the value for buckling must equal the first eigenvalue the following holds true.

$$\frac{k_\infty - 1}{L^2} = B_1^2$$

Note: The important concept here is that the left side of this equation is determined by the material properties of the system. The right hand side of the equation is determined by the geometry and dimensions. The two together determine the requirements for criticality.

The subscript indicating the first eigenvalue is usually dropped. Rearranging the equation into the form most often seen gives the following.

$$\frac{k_\infty}{1 + L^2B^2} = 1$$

This is the one-group critical equation for a bare reactor.

miscellaneous information

item	name	value
N_{fe}	atom density	0.848×10^{24} atoms/cm³
N_{pu}	atom density	0.0493×10^{24} atoms/cm³
L^2	diffusion area	D/Σ_a

[9] The diffusion equation which started this derivation assumes a bare reactor (i.e., a reactor without a blanket or a reflector). See the Reference section for a text defining nuclear terms.

[10] The one-group diffusion equation is solved to determine the necessary flux equation (i.e., the flux shape). The buckling must equal the first eigenvalue squared in order to meet boundary conditions. The first eigenvalue is the only one of concern in a critical reactor. See the Reference section texts for a more complete explanation.

NUCLEAR THEORY–10

A homogeneous mixture of Fe-55 and Pu-239 is to be made into the shape of a bare sphere. Estimate the critical radius of this fast reactor?

(A) 2 cm
(B) 6 cm
(C) 12 cm
(D) 66 cm

NUCLEAR THEORY–11

What is the non-leakage probability for the bare reactor in the Prob. 10?

(A) 0.1
(B) 0.3
(C) 0.4
(D) 0.6

NUCLEAR THEORY–12

What is the infinite multiplication factor for a thermal reactor with a thermal utilization of 0.5? Use typical parameter values.

(A) 1.0
(B) 1.5
(C) 2.0
(D) 2.5

NUCLEAR THEORY–13

For a spherical thermal reactor with a water moderator in a homogeneous mixture, what is the value of the infinite multiplication factor necessary for the reactor to be critical? The radius of the reactor is 1 m. Use the modified one-group critical equation.

(A) 1.0
(B) 1.5
(C) 2.0
(D) 2.5

Nuclear Theory (NT) problems 14 and 15 are based on the following information and equations.

parameters of a critical bare cylindrical reactor

buckling	flux
$\left(\frac{2.405}{R}\right)^2+\left(\frac{\pi}{H}\right)^2$	$AJ_0\left(\frac{2.405r}{R}\right)\cos\left(\frac{\pi z}{H}\right)$
constant A	
$\frac{3.63P}{VE_R\Sigma_f}$	

R = radius of the cylinder
r = position from axis center
H = height of core
z = position along the axis ($z = 0$ at $H = 0.5$)
J_0 = Bessel function of the first kind of order zero
P = power
V = volume of core
E_R = recoverable energy per fission (200 MeV)
Σ_f = macroscopic fission cross section

When the thermal flux is the desired quantity, adjustments to the given formulas must be made. [11]

A certain bare cylindrical reactor with a radius of 0.20 m and a height of 1 m consists of a homogeneous mixture of U-235 and ordinary water with a density of 1 g/cm^3.

Note: Ignore density and temperature corrections (i.e., use 20°C values).[12] The non-$1/v$ factor for water may be taken as equal to one.

[11] When the thermal flux is the desired quantity, instead of the 2200 m/s flux, the average macroscopic fission cross section must be used.

$$\bar{\Sigma}_f = N\bar{\sigma}_f = N\left(\frac{\sqrt{\pi}}{2}\right)\left(g_f(t)\right)\left(\sigma_f(E_0)\right)\left(\sqrt{\frac{T_0}{T}}\right)$$

This includes a term which accounts for the ratio of the 2200 m/s flux ϕ_0 to the thermal flux ϕ_{th}; a non-$1/v$ factor; the 2200 m/s microscopic cross section; and a temperature adjustment if not at 20°C (i.e., 293.61 K). All the macroscopic cross sections should be handled in this manner when used with the thermal flux.

[12] This assumption is made primarily for simplicity. Also, this problem could represent uranium waste mixed with water in a barrel. Thus, the actual temperature is approximately the reference temperature of 20°C (i.e., 293.61 K). Normally the thermal diffusion parameters, the thermal diffusion coefficient and the thermal diffusion area, are adjusted for the density and temperature of the mixture. See the Reference section for additional information.

NUCLEAR THEORY–14

What is the *critical* mass? Use modified one-group theory.

(A) 150 g
(B) 240 g
(C) 1550 g
(D) 4800 g

NUCLEAR THEORY–15

What is the flux at the center of the reactor (i.e., where the radius r = 0) at the core mid-plane (i.e., where z = 0) if the output power is 100 kW?

(A) 1.8×10^{12} $cm^{-2}\cdot s^{-1}$
(B) 5.6×10^{12} $cm^{-2}\cdot s^{-1}$
(C) 13.0×10^{12} $cm^{-2}\cdot s^{-1}$
(D) 13.6×10^{12} $cm^{-2}\cdot s^{-1}$

NUCLEAR THEORY–16

A reflector has what effect(s)?

(A) flattens the thermal flux
(B) flattens the fast flux
(C) reduces the critical mass
(D) A and C

NUCLEAR THEORY–17

The reactor period T for an infinite thermal reactor, without delayed neutrons, is given by

$$T = \frac{\ell_p}{k_\infty - 1}$$

ℓ_p = prompt neutron lifetime[13]

If the reactivity ρ changes by 0.1% in a critical thermal reactor operating at 100 MW, what is the power one second later?

(A) 110 MW
(B) 400 MW
(C) 2×10^6 MW
(D) 3×10^{45} MW

NUCLEAR THEORY–18

The reactivity in a reactor fueled with U-235 changes by 0.2% due to an outward rod shim. What is the reactivity change in cents?

(A) 2 cents
(B) 31 cents
(C) 77 cents
(D) 102 cents

[13] The value for the prompt lifetime is approximately 10^{-4} s in water moderated thermal reactors and 10^{-7} s in fast reactors.

NUCLEAR THEORY–19

Consider the following table of differential rod worth and integral rod worth.

DRW $\Delta\rho/\rho$	IRW $\rho(x)/\rho(H)$ \$	rod position (H) (arbitrary units)
0	0	0
7.48×10^{-3}	1.15	25
4.00×10^{-3}	6.15	50
7.35×10^{-2}	11.30	75
8.00×10^{-2}	12.30	100

DRW = differential rod worth
IRW = integral rod worth

The delayed neutron fraction β is taken to be 0.0065. It takes three minutes for a rod to move from the bottom to the top of the core. Safety considerations and regulations limit the reactivity insertion rate to $14 \times 10^{-4}\ \Delta\rho/\rho$ or approximately 21 cents per second during steady state conditions.

Consider the following curves of rod worth.

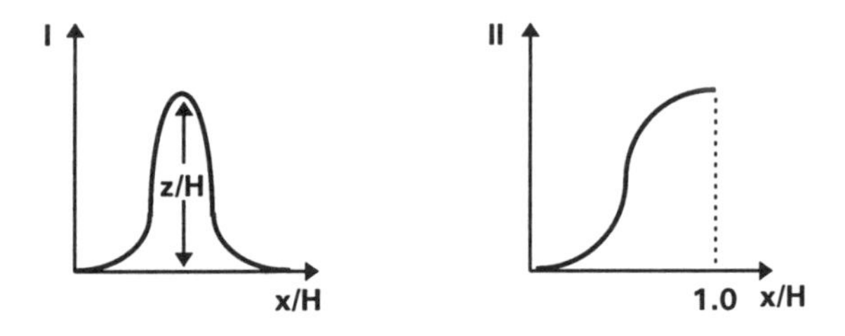

What is the maximum reactivity insertion rate? Which curve shown represents the shape of a differential rod worth curve?

(A) 3 cents/s I
(B) 9 cents/s II
(C) 12 cents/s I
(D) 42 cents/s II

NUCLEAR THEORY–20

The concentration of Xe-135, a fission product poison, is given by the following.

$$\frac{dN_{\text{Xe}}}{dt} = \lambda_I N_I + \gamma_{\text{Xe}} \overline{\Sigma}_f \phi_{\text{TH}} - \lambda_{\text{Xe}} N_{\text{Xe}} - \overline{\Sigma}_{a,\text{Xe}} \phi_{\text{TH}}$$

N_{se} = atom density of xenon
λ_I = decay constant of iodine
N_I = atom density of iodine
γ_{Xe} = fission yield of xenon
Σ = macroscopic cross section
ϕ_{TH} = thermal flux

Consider the following graph of xenon concentration over time.

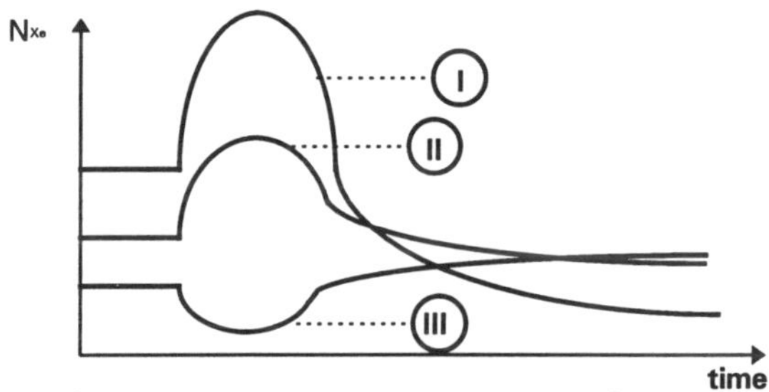

Which curve represents an up-power, down-power, and shutdown transient?

	up	down	shutdown
(A)	I	II	III
(B)	II	III	I
(C)	III	I	II
(D)	III	II	I

5 NUCLEAR INSTRUMENTS

NUCLEAR INSTRUMENTS–1

The general method of producing x-rays from target material for use in radiography or for medical purposes is shown.

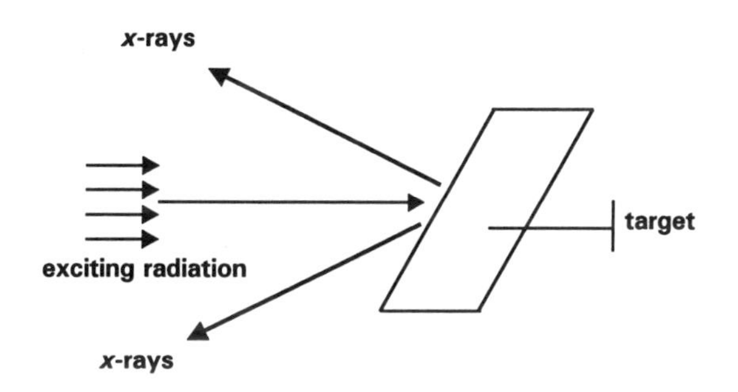

The exciting radiation can be an ionizing radiation.[1] Assume excitation of the target results in the expulsion of a K-shell electron which is then filled by an L-shell electron as shown.[2]

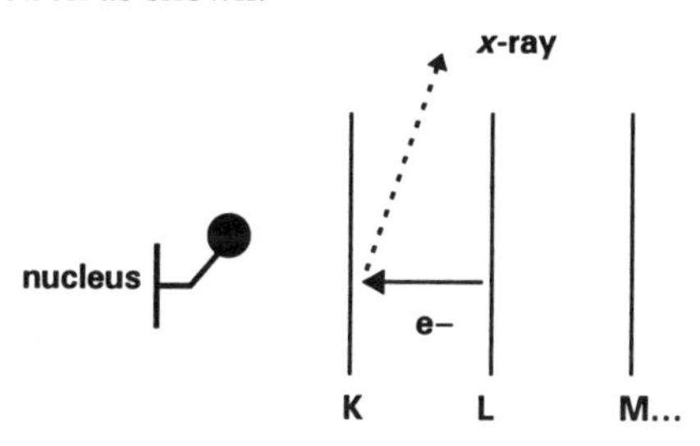

The emitted x-ray is called by what term?

(A) Auger x-ray
(B) internal conversion x-ray
(C) Bremsstrahlung x-ray
(D) K_α characteristic x-ray

[1] The range under discussion is from approximately 10 eV, the minimum energy required to cause ionization in typically used materials, to about 20 MeV which is the upper bound of energy for concern in nuclear science and technology.
[2] The labels K, L, M, and so on represent the total quantum numbers 1, 2, 3, and so on. The designations s, f, p, d, and so on are representative of angular-momentum states.

NUCLEAR INSTRUMENTS–2

Neutron sources are required for the testing of nuclear instrumentation and for the calibration of neutron radiation detectors. Two significant radioisotope photoneutron sources follow.

$$^{9}_{4}\text{Be} + h\nu \rightarrow {}^{8}_{4}\text{Be} + {}^{1}_{0}\text{n}$$

$$^{2}_{1}\text{H} + h\nu \rightarrow {}^{1}_{1}\text{H} + {}^{1}_{0}\text{n}$$

The Q value for the beryllium reaction is –1.666 MeV. The Q value for the hydrogen reaction is –2.226 MeV.

If the beryllium source is used, what type of radiation, based on the frequency required, is necessary to cause the reaction to occur?

(A) radio
(B) infrared
(C) visible
(D) gamma

NUCLEAR INSTRUMENTS–3

A source and background count of 760 counts over a three minute period is measured by a laboratory detector. A stable average background count taken over a long period of time measures 100 counts per minute (cpm).

What is the net count rate and standard deviation for the source alone?

(A) 150 cpm ± 9 cpm
(B) 660 cpm ± 9 cpm
(C) 150 cpm ± 13 cpm
(D) 660 cpm ± 13 cpm

NUCLEAR INSTRUMENTS–4

A 100 cm^2 swipe is taken from an area suspected of harboring surface contamination. The result of an initial one minute count is

125 counts

Due to the unexpectedly high results, two additional counts are completed of three minutes duration. The results are as follows.

350 counts
362 counts

The one minute background count completed just prior to the survey indicated 20 counts.

A report is to be completed based on this information. What is the net average count rate per minute and the standard deviation in counts per minute (cpm)?

(A) 100 cpm ± 13 cpm
(B) 120 cpm ± 12 cpm
(C) 120 cpm ± 13 cpm
(D) 360 cpm ± 13 cpm

NUCLEAR INSTRUMENTS–5

Three statistical models used in nuclear engineering are the binomial, poisson, and gaussian distributions. The first is used in constant probability processes; the second where the probability of success is small (e.g., long half-life, low observation time); the third where the probability of success is large (e.g., 20 or more counts).

If the definition of success is the decay of a nucleus during an observation, what is the probability of success during a five minute observation of a radioactive source with a half-life of 53 minutes?

(A) 0.06
(B) 0.09
(C) 0.91
(D) 0.94

NUCLEAR INSTRUMENTS–6

A simplified detector and output circuitry is shown.

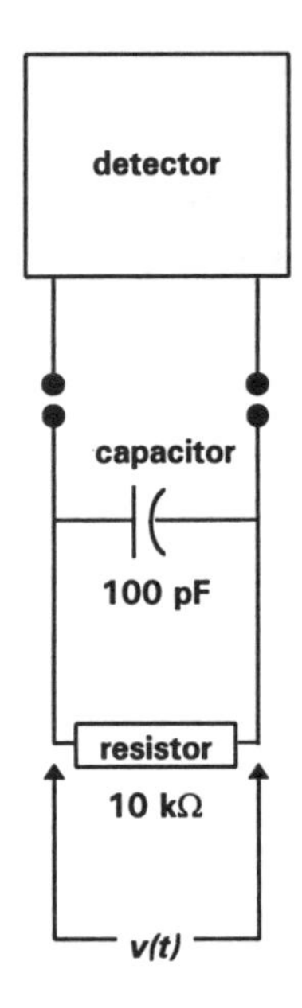

If the time constant λ is much larger than the detector charge collection time (i.e., a large value of RC), the output signal for a single interaction in the detector is as follows.

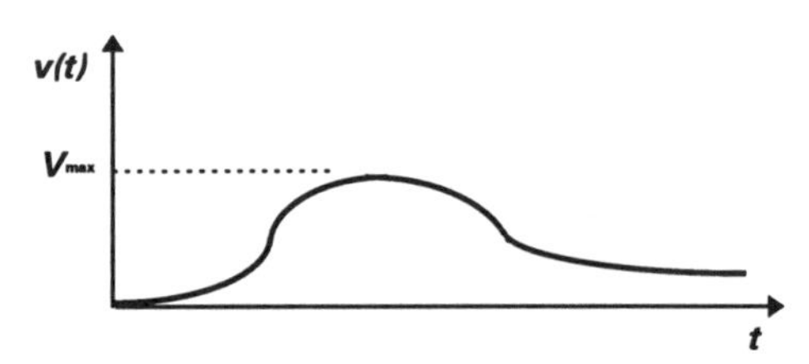

Assume one interaction in the detector creates 10^5 electrons of charge. What mode is the detector operating in and what is the approximate value of V_{max}?

(A) mean square voltage; 10 μV
(B) Campbelling; 0.02 V
(C) current; 10 μV
(D) pulse; 0.02 V

NUCLEAR INSTRUMENTS–7

Iron-59 is believed present in a sample. To confirm, a radiation detector is set to measure the energy of the two gammas emitted during the decay expected. The gammas have an approximate energy of 1.30 MeV and 1.10 MeV.

What resolution is required by the detector to ensure these gamma energies can be detected as separate entities?

(A) 2%
(B) 10%
(C) 17%
(D) 65%

NUCLEAR INSTRUMENTS–8

A small DC ion chamber device known as pocket chamber or pocket dosimeter is often used to monitor exposure. They are fitted with an integral fiber electroscope which can be read when held up to a light. The initial charge zeroes the scale. The scale moves when exposed to a radiation field which results in a decrease of voltage in the chamber. A generic setup for such an instrument is shown.

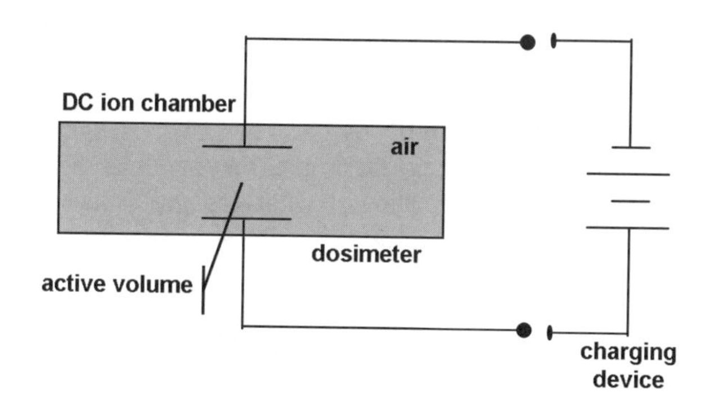

The active volume is $15\ \text{cm}^3$. The capacitance of the device is 75 pF. The battery voltage is 6 V. The air is at STP.[3]

What gamma ray exposure will reduce the initial chamber voltage to 4 V?

(A) 600 μR
(B) 30 mR
(C) 1 R
(D) 100 R

NUCLEAR INSTRUMENTS–9

The characteristics of gas-filled detectors operated in the pulse mode can represented by the following graph.

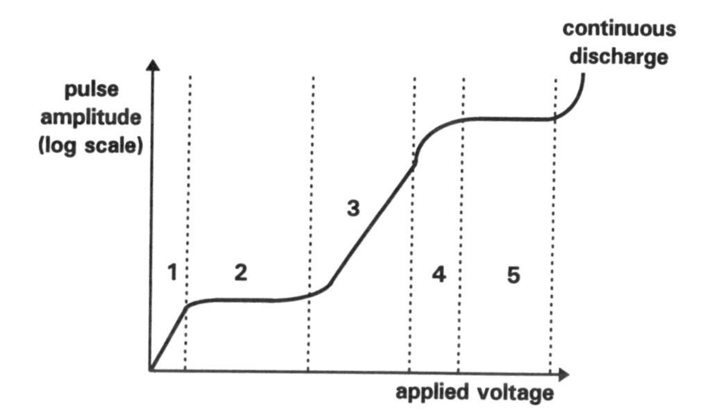

The regions under consideration are numbered 1–5. Consider now the following characteristics.

designation	characteristics
I	DC ion chamber region of operation
II	true proportionality; conventional proportional counters region of operation
III	recombination of created ion pairs occurs before collection
IV	space charge effects can come into play altering the electric field of the detector
V	a single ionizing event saturates the detector

Match the characteristic with the correct region.

(A) I 1 II 3 III 4 IV 5 V 2
(B) I 2 II 3 III 1 IV 4 V 5
(C) I 2 II 4 III 1 IV 3 V 5
(D) I 5 II 4 III 3 IV 2 V 1

[3] The definition of STP, standard temperature and pressure, actually varies. The one most encountered in physics is 0°C and one atmosphere (i.e., 760 torr). The density of one cubic centimeter of dry air is then 0.001293 g.

Nuclear Instruments (NI) problems 10–13 are based on the following information, drawing, and graph.

Crystalline sodium iodide with a trace of thallium iodide NaI[Tl] is the most common material for scintillation spectrometry.[4] Radiation interacts with the material exciting the individual atoms of the structure. They undergo scintillation producing a flash of light.[5] This light is coupled directly or via light pipes to a photomultiplier (PM) tube.[6] A photocathode sensing the weak scintillation light signal, produces free electrons which are then amplified in number in a multiplier section.

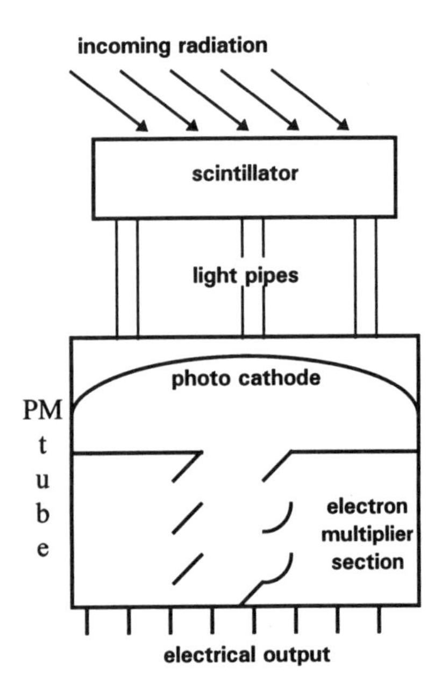

The electrical output, which goes to a multichannel analyzer (MCA), is shown on the next graph.[7] Numerous properties are included.[8] Different size detectors with multiple energy ranges are shown on the same graph.

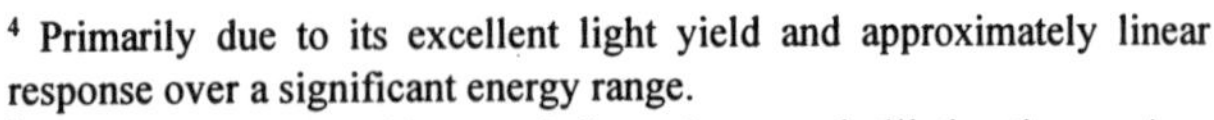

[4] Primarily due to its excellent light yield and approximately linear response over a significant energy range.

[5] Numerous texts provide more information on scintillation theory. As a nuclear engineer, the resulting output is of more immediate concern.

[6] Semiconductor photodiodes are also used.

[7] Several types of electrical circuitry exist between the detector and the multichannel analyzer (MCA). See the Reference section for more information.

[8] Different types of detectors are available, such as lithium-drifted silicon detectors; nevertheless, the resulting output is generally the same shape and the principles are the similar. What differs is the radiation interaction in the detector itself.

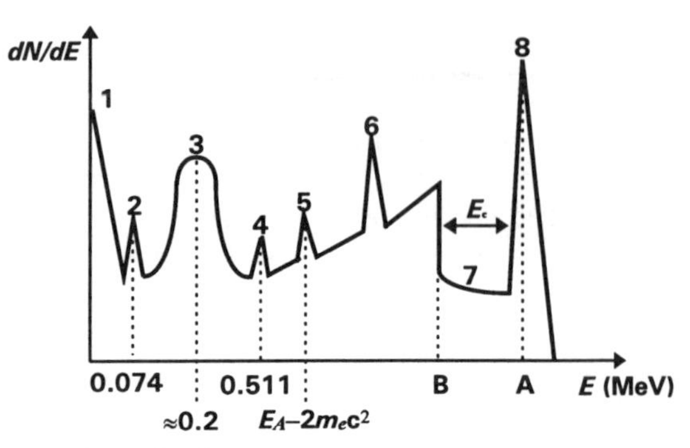

N = number of pulses
E = energy

NUCLEAR INSTRUMENTS–10

A given photocathode gain is 10^6 and the electron output pulse from the PM tube occurs in 6 ns.[9] What is the approximate peak pulse current output of the detector if incoming radiation causes a scintillation event releasing 1100 photoelectrons from the photocathode?

(A) 0.03 mA
(B) 0.30 mA
(C) 3.00 mA
(D) 30.0 mA

NUCLEAR INSTRUMENTS–11

One of the gammas emitted by N-16 has a wavelength of 1.75×10^{-3} Å. Assuming these are the gammas being detected by the graph shown, what are the values of points A and B?

(A) E_A 6.1 MeV E_B 2.10 MeV
(B) E_A 6.1 MeV E_B 5.84 MeV
(C) E_A 7.1 MeV E_B 3.10 MeV
(D) E_A 7.1 MeV E_B 6.84 MeV

[9] Both the gain and the pulse width in units of time could be taken from the manufacturer's specifications for the PM tube.

NUCLEAR INSTRUMENTS–12

In the detector (i.e., the scintillator) the following events can occur.

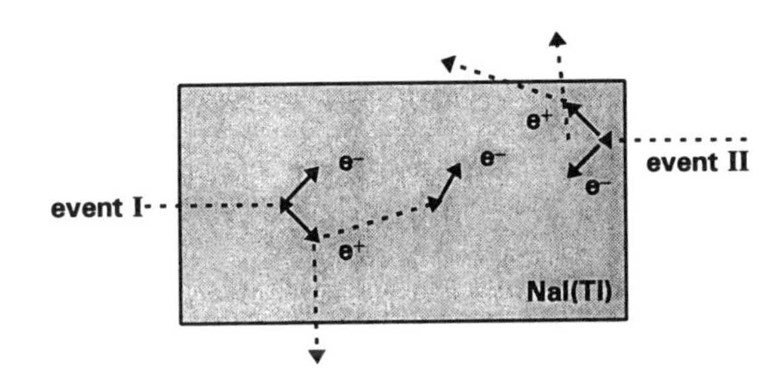

Assuming the event I gamma possesses the minimum energy necessary, how much energy is deposited in the detector? Event II represents what region on the graph?

(A) 0.256 MeV; 5
(B) 0.511 MeV; 5
(C) 0.765 MeV; 1
(D) 1.02 MeV; 1

NUCLEAR INSTRUMENTS–13

In a particular application, a scintillation detector is surrounded by lead as shown in order to measure the response of a radioactive source.

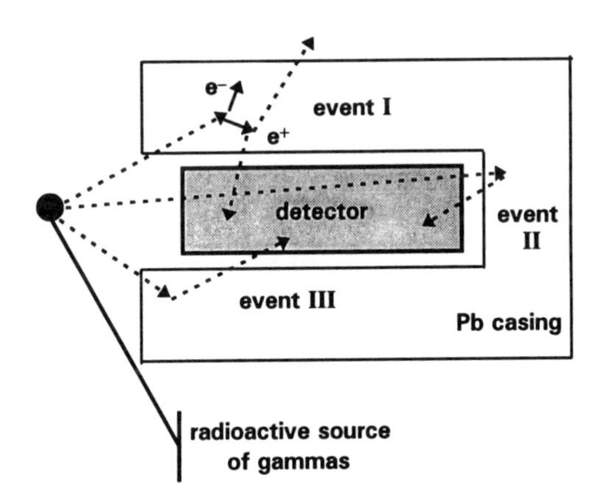

Which event is responsible for region 2 on the graph? for region 3?

(A) region 2: I region 3: III
(B) region 2: II region 3: III
(C) region 2: II region 3: I
(D) region 2: III region 3: II

NUCLEAR INSTRUMENTS–14

Scintillation and Geiger-Mueller detector charge outputs are often large enough to be used directly in the follow on electronic circuitry. Most other types of detectors require signal amplification. For that purpose, a simple preamplifier is shown.

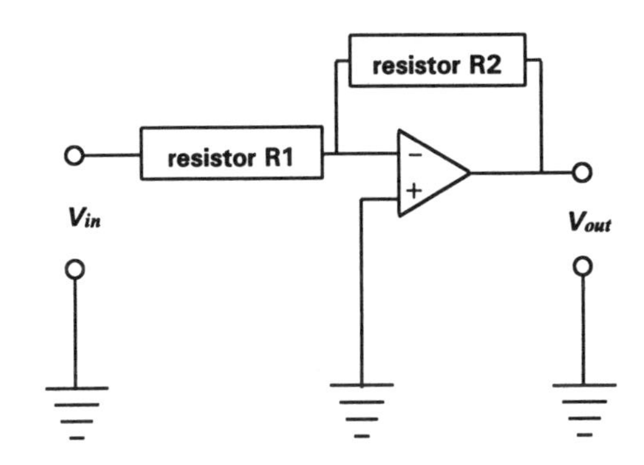

The gain of the amplifier is 10^5. The values of resistors R1 and R2 are 100 Ω and 10 kΩ respectively.

If the input voltage is 3 mV, what is the output voltage?

(A) −0.30 V
(B) −3.00 V
(C) −30.0 V
(D) −300 V

NUCLEAR INSTRUMENTS–15

The detection efficiency for neutrons of a certain BF_3 tube with a boron density of 2×10^{19} cm^{-3} is given by[10]

$$eff(E) = 1 - \exp\left(-\Sigma_a(E)L\right)$$

$\Sigma_a(E)$ = macroscopic absorption cross section of B-10 at energy E
L = active length of the tube

The approximate ranges in which nuclear radiation detectors used as power sensors are shown.[11]

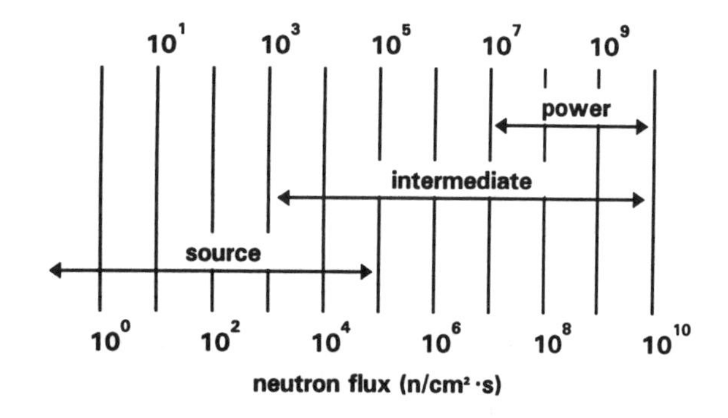

How long must the active tube length be to get at least 90% efficiency for thermal neutrons? In what range would this detector be used?

(A) 30 cm; power range
(B) 60 cm; intermediate range
(C) 30 cm; source range
(D) 60 cm; source range

[10] This is an approximate relationship and varies with the tube pressure, construction details, and neutron angle of incidence.
[11] The flux indicated is that which exists at the detector location outside the core.

6 NUCLEAR ISSUES

NUCLEAR ISSUES–1

Which of the following are important issues and concerns for the nuclear industry in the future?

(A) Kyoto protocol
(B) DOE nuclear waste policy
(C) NRC licensing and decommissioning policy
(D) all of the above

7 NUCLEAR POWER SYSTEMS: SOLUTIONS

Nuclear Power Systems (NPS) problems 1–10 are based on the following information and illustration.

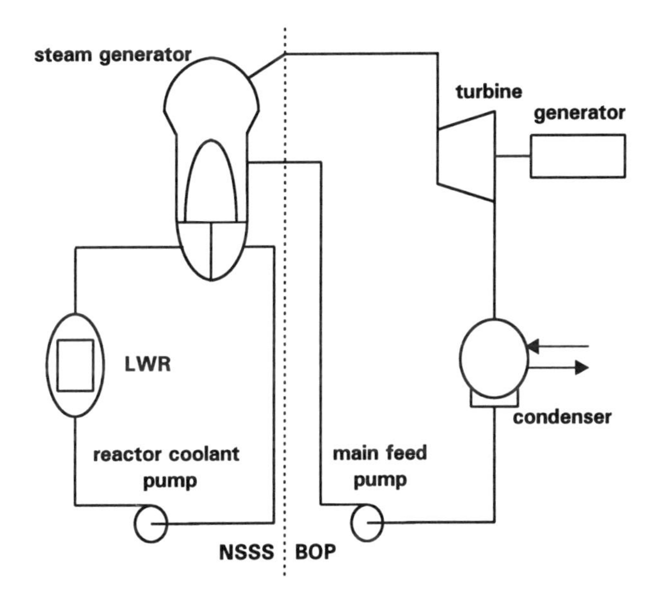

Terminology
BOP = balance of the plant
LWR = light water reactor
NSSS = nuclear steam supply system

Plant Data
active core height = 3.5 m
total fuel loading = 90 000 kg
UO_2 density = 10 g/cm^3
enrichment = 3 w/o [weight percent]
150 fuel assemblies
200 fuel pins per assembly
electrical generating capacity = 1000 MW
plant thermal efficiency (η_{th}) = 30%
plant availability = 75%

Scientific Data
E_r = energy recoverable per fission = 200 MeV
E_d = energy deposited in fuel per fission = 2.75×10^{-11} J
$\bar{\sigma}_{f,25}$ = average microscopic cross section for fission in U-235 = 366 barns[1,2]
c_p = specific heat capacity of water at 20°C (68°F) = 4.187 J/g·°C (1.00 Btu/lbm-°F)

NUCLEAR POWER SYSTEMS–1

What is the approximate thermal power output (P_{th}) of the core?

(A) 300 MW
(B) 1000 MW
(C) 1300 MW
(D) 3300 MW

Solution:

The formula for the thermal power (P_{th}) is the output power (P_{out}) divided by the thermal efficiency (η_{th}). In terms of an equation,

$$P_{th} = \frac{P_{out}}{\eta_{th}}$$

The electrical generating capacity is the P_{out} term.

---

[1] The value shown is obtained by using the cross section value for U-235 tabulated for a neutron velocity of 2200 m/sec (the thermal value); averaging by assuming a Maxwellian distribution of neutron velocities (i.e., dividing by 1.128); multiplying by a non-1/v correction factor which is temperature dependent and also routinely tabulated; and finally, compensating for temperature of the uranium fuel as the above neutron cross sections are most often expressed at a temperature of 20°C. Also note that the symbols for the cross section vary from text to text (e.g., σ, $\bar{\sigma}$, $\bar{\sigma}'$, or $\hat{\sigma}$) as well as others, depending upon which of the operations mentioned are complete and the preferred symbology. See the Nuclear Theory section for additional information.

[2] The unit barns is sometimes symbolized with the letter b.

Substituting gives

$$P_{th} = \frac{1000\ \text{MW}}{0.30}$$

$$= \boxed{3333\ \text{MW}\quad (3300\ \text{MW})}$$

Answer is D.

NUCLEAR POWER SYSTEMS–2

What is the atom density of U-235 (N_{25}) in the core?

(A) 5.9×10^{20} cm-3
(B) 6.7×10^{20} cm-3
(C) 6.8×10^{20} cm-3
(D) 7.6×10^{20} cm-3

Solution:

The applicable formula is[3]

$$N_{25} = \frac{\rho N_A}{\text{MW}}$$

ρ = density
N_A = Avogadro's number
MW = molecular weight

Thus,

$$N_{25} = \frac{\left(10\,\frac{\text{g}}{\text{cm}^3}\right)\left(6.02\times 10^{23}\,\frac{\text{molecules UO}_2}{\text{mol}}\right) \times \left(\frac{1\ \text{atom U}}{1\ \text{molecule UO}_2}\right)\left(\frac{0.03\ {}^{235}\text{U}}{1\ \text{atom U}}\right)}{\left(\left(238\ \text{g}\ {}^{238}\text{U} + (2)(16\ \text{g}\ {}^{16}\text{O})\right)\frac{\text{g}}{\text{mol}}\right)}$$

The 0.03 is the enrichment in weight percent and represents the fraction of all the uranium which is of the fissile U-235 form.[4] Solve and cancel units.[5]

$$N_{25} = \boxed{6.7\times 10^{20}\ \text{atoms}\ {}^{235}\text{U}/\text{cm}^3}$$

Answer is B.

NUCLEAR POWER SYSTEMS–3

What is the average macroscopic fission cross section for the U-235 in the core?

(A) 2.16×10^{-1} cm-1
(B) 2.45×10^{-1} cm-1
(C) 2.49×10^{-1} cm-1
(D) 2.78×10^{-1} cm-1

Solution:

In general, any macroscopic cross section can be found from the following.[6]

$$\Sigma = N\sigma$$

The appropriate designations must be added to each term.

$$\overline{\Sigma}_{f,25} = N_{25}\overline{\sigma}_{f,25}$$

The density (N_{25}) of the U-235 was solved for in Prob. 2 and the average microscopic cross section for U-235 is given information. Thus,

$$\overline{\Sigma}_{f,25} = \left(6.69\times 10^{20}\ \frac{\text{atoms}}{\text{cm}^3}\right)(366\ \text{barns})\left(\frac{10^{-24}\ \text{cm}^2}{1\ \text{barn}}\right)$$

$$= \boxed{2.45\times 10^{-1}\ \text{cm}^{-1}}$$

Answer is C. [7]

[3] While the formula could be determined by referring to an appropriate text, it's instructive to note that many desired quantities, including this one, can be determined using unit analysis.
[4] Fissile nuclei are those that lead to fission following the absorption of a neutron with zero kinetic energy.
[5] "Atoms" and "molecules" are not units requiring canceling and are shown for clarification only, in this and all other problems.
[6] The microscopic cross section can be thought of as a measure of the probability of interaction with a given nucleus while the macroscopic cross section represents the probability per unit path length of an interaction in the material of concern (i.e., the material represented by the density term, N). The Nuclear Theory section has more information.
[7] Similar calculations of a scoping or preliminary nature are often carried out without the corrections mentioned in a previous footnote and using a value for the cross section taken directly from a table (i.e., referenced to 20°C). In this case the average macroscopic fission cross section becomes the fission cross section. Usually, this means the average bar is removed from the Σ. Be aware, however, the use of the average bar is not universal.

NUCLEAR POWER SYSTEMS–4

What is the average thermal neutron flux in the core at the beginning of the first cycle?[8]

(A) 8.81×10^{11} neutrons/cm²·s
(B) 2.93×10^{13} neutrons/cm²·s
(C) 4.72×10^{13} neutrons/cm²·s
(D) 3.00×10^{16} neutrons/cm²·s

Solution:

$$P_{\text{th}} = E_r \overline{\Sigma}_{f,25} \overline{\varphi}_{25} V$$

V = volume of the core

Rearranging,

$$\overline{\varphi}_{25} = \frac{P_{\text{th}}}{E_r \overline{\Sigma}_{f,25} V}$$

The thermal power P_{th} was found in Prob. 1, E_r is given information, and the average $\Sigma_{f,25}$ was found in Prob. 3. Therefore, only the volume, V, need be found to solve for the desired quantity. The volume is given by

$$V = \frac{m}{\rho}$$

m = mass
ρ = density

Thus,

$$V = \frac{(90\,000 \text{ kg fuel})\left(\frac{10^3 \text{ g}}{1 \text{ kg}}\right)}{10 \frac{\text{g}}{\text{cm}^3}} = 9.00 \times 10^6 \text{ cm}^3$$

[8] The beginning of first cycle was specified since the atom density of uranium 235 decreases with core operation and the average thermal neutron flux increases. See the Nuclear Theory section for more information.

Substituting into the equation for the thermal flux gives

$$\overline{\varphi}_{25} = \frac{P_{\text{th}}}{E_r \overline{\Sigma}_{f,25} V}$$

$$= \left(\frac{(3333 \text{ MW})\left(\frac{10^6 \text{ W}}{1 \text{ MW}}\right)\left(\frac{1 \text{ J}}{\text{W}\cdot\text{s}}\right)}{(200 \text{ MeV})\left(10^6 \frac{\text{eV}}{\text{MeV}}\right)\left(1.6 \times 10^{-19} \frac{\text{J}}{\text{eV}}\right)}\right) \times \left(\frac{1}{(2.45 \times 10^{-1} \text{ cm}^{-1})(9.00 \times 10^6 \text{ cm}^3)}\right)$$

$$= \boxed{4.72 \times 10^{13} \text{ neutrons/cm}^2 \cdot \text{s}}$$

Answer is C.

NUCLEAR POWER SYSTEMS–5

The core is expected to operate at 80% of full power the majority of the time. (Assume the fuel is uniformly distributed throughout the core.) Given this power level, what is most nearly the average thermal power density, P_d, in the fuel?

(A) 88 W/cm³
(B) 296 W/cm³
(C) 370 W/cm³
(D) 300×10^3 W/cm³

Solution:

The thermal power density at 80% is given by

$$P_d = \frac{0.80 P_{\text{th}}}{V}$$

The thermal power output of the core at full load is

$$P_{\text{th}} = \frac{P_{\text{out}}}{\eta_{\text{th}}} = \frac{1000 \text{ MW}}{0.30} = 3333 \text{ MW}$$

The volume of the core is

$$V = \frac{m}{\rho} = \frac{(90\,000\ \text{kg fuel})\left(\frac{10^3\ \text{g}}{1\ \text{kg}}\right)}{10\ \frac{\text{g}}{\text{cm}^3}} = 9.00 \times 10^6\ \text{cm}^3$$

Substituting into the thermal density equation gives

$$P_d = \frac{0.80 P_{th}}{V} = \frac{(0.80)(3\,333\ \text{MW})\left(\frac{10^6\ \text{W}}{1\ \text{MW}}\right)}{9.00 \times 10^6\ \text{cm}^3} = \boxed{296\ \text{W/cm}^3}$$

Answer is B.

NUCLEAR POWER SYSTEMS–6

What is the mass of U-235 in the core?

(A) 2310 kg
(B) 2340 kg
(C) 2350 kg
(D) 2380 kg

Solution:

The atom density (N_{25}) in the core was found in Prob. 2.

$$N_{25} = \frac{\rho N_A}{\text{MW}} = 6.7 \times 10^{20}\ \text{atoms}\ ^{235}\text{U/cm}^3$$

Note that density is

$$\rho = \frac{m}{V}$$

The volume was calculated to be $9.00 \times 10^6\ \text{cm}^3$ in Prob. 5.

Substituting the formula for density into the equation for N_{25} gives

$$N_{25} = \frac{\left(\frac{m}{V}\right) N_A}{\text{AW}}$$

(The atomic weight (AW) replaces the molecular weight (MW) because the formula as now used concerns only atoms of U-235 instead of molecules of UO_2 as in Prob. 2.)

Rearranging, solving for the mass (m), and substituting the previously known values from Probs. 2 and 5 gives

$$m = \frac{N_{25}\text{AW}V}{N_A} = \frac{\left(6.7 \times 10^{20}\ \frac{\text{atoms}\ ^{235}\text{U}}{\text{cm}^3}\right)\left(235\ \frac{\text{g}}{\text{mol}}\right)\left(\frac{1\ \text{kg}}{10^3\ \text{g}}\right) \times \left(9.00 \times 10^6\ \text{cm}^3\right)}{6.02 \times 10^{23}\ \frac{\text{atoms}\ ^{235}\text{U}}{\text{mol}}}$$

$$= \boxed{2354\ \text{kg}\ \ (2350\ \text{kg})}$$

Answer is C.

NUCLEAR POWER SYSTEMS–7

Assume that during operation an average thermal neutron flux of 3.00×10^{13} neutrons/cm^3·s exists. The cycle for this plant is to be one year of operation followed by refueling of one-third of the core. What is the *fluence* (Φ) during one cycle?[9]

(A) 9.45×10^{13} neutrons/cm^2
(B) 3.15×10^{20} neutrons/cm^2
(C) 4.73×10^{20} neutrons/cm^2
(D) 9.46×10^{20} neutrons/cm^2

[9] This information would be of concern to materials engineers and others attempting to determine the radiation effects on various portions of the plant. See the Nuclear Radiation section for more information.

Solution:

The fluence is defined as the time integrated flux per unit area. In mathematical terms,

$$\Phi = \int_0^T \varphi(t)\, dt$$

T = the cycle period

In this case, the flux (φ) is averaged. Therefore, the flux φ (t) = constant = 3.00×10^{13} neutrons/cm^2·s. Thus, the fluence formula can be rearranged to give

$$\begin{aligned}\Phi &= \overline{\varphi} \int_0^T dt = \overline{\varphi}\, T \\ &= \left(3.00 \times 10^{13} \frac{\text{neutrons}}{\text{cm}^2 \cdot \text{s}}\right) \\ &\quad \times (1\ \text{yr})\left(\frac{365\ \text{days}}{1\ \text{yr}}\right)\left(\frac{24\ \text{hr}}{1\ \text{day}}\right)\left(\frac{3600\ \text{s}}{1\ \text{hr}}\right) \\ &= \boxed{9.46 \times 10^{20}\ \text{neutrons/cm}^2}\end{aligned}$$

Answer is D.

NUCLEAR POWER SYSTEMS–8

What is the fuel pin average thermal lineal power density (P_l)?

(A) 0.87 kW$_{th}$/ft
(B) 2.90 kW$_{th}$/ft
(C) 3.77 kW$_{th}$/ft
(D) 9.66 kW$_{th}$/ft

Solution:

First, define the lineal power density (P_l) as follows.

$$P_l = \frac{\text{average kW}_{\text{th}} \text{ in a fuel pin}}{\text{height of the fuel pin in feet}}$$

The average kW$_{th}$ in a fuel pin can be determined from the following.

$$\begin{aligned}\overline{\text{kW}}_{\text{th,pin}} &= \frac{P_{\text{th}}}{\text{no. of fuel pins}} \\ &= \frac{(3333\ \text{MW})\left(\dfrac{10^3\ \text{kW}}{1\ \text{MW}}\right)}{(150\ \text{fuel assemblies})\left(200\ \dfrac{\text{fuel pins}}{\text{fuel assembly}}\right)} \\ &= 111.1\ \text{kW}_{\text{th}}/\text{fuel pin}\end{aligned}$$

The *active core height*[10] is the height of any given fuel pin, assuming they are all the same height—a reasonable assumption. Thus, fuel pin height is given by

$$\begin{aligned}\text{fuel pin height} &= \text{active core height} \\ &= (3.5\ \text{m})\left(\frac{3.28\ \text{ft}}{1\ \text{m}}\right) \\ &= 11.5\ \text{ft}\end{aligned}$$

Substituting the calculated values into the definition for P_l gives

$$\begin{aligned}P_l &= \frac{\text{average kW}_{\text{th}} \text{ in a fuel pin}}{\text{height of the fuel pin in feet}} \\ &= \frac{111.1\ \dfrac{\text{kW}_{\text{th}}}{\text{fuel pin}}}{11.5\ \text{ft}} \\ &= \boxed{9.66\ \text{kW}_{\text{th}}/\text{fuel pin-ft}\quad (9.66\ \text{kW}_{\text{th}}/\text{ft})}\end{aligned}$$

Answer is D.

[10] The fuel pins could actually be much larger than the active core height. During operation not all of the fuel is exposed (i.e., in an area removed from control rods) or "active." Since no additional information was given, assuming the active core height and the fuel pin height are identical is reasonable in this case.

NUCLEAR POWER SYSTEMS–9

A certain reactor's heat sink (i.e., the cooling water used in its condenser) is a large lake. State regulations require that any water discharge to the lake be no more than 2°F greater than the mean temperature of the lake. Given this restriction, what mass flow rate ($\dot{m}$) must exist in the condenser?

(A) 1.71×10^9 lbm/hr
(B) 3.98×10^9 lbm/hr
(C) 5.63×10^9 lbm/hr
(D) 6.83×10^9 lbm/hr

Solution:

The heat transfer rate of the condenser representing the heat rejected to the atmosphere by the condenser is given by

$$\dot{Q}_{\text{cond}} = \dot{m} c_p \Delta T$$

ΔT = the temperature difference between the outgoing and incoming condenser water

Rearranging gives

$$\dot{m} = \frac{\dot{Q}_{\text{cond}}}{c_p \Delta T}$$

Recall that the thermal power output of the plant, P_{th}, is 3333 MW and the output power, P_{out}, is 1000 MW. [11,12] Thus,

$$\dot{Q}_{\text{cond}} \approx 3333\,\text{MW} - 1000\,\text{MW} = 2333\,\text{MW}$$

[11] The thermal power may be represented as the total heat rejected, Q_{rej}, by the plant in some problems. Do not confuse this with the heat "rejected" by the condenser, Q_{cond}, which in some texts is written as Q_{rej}. Obviously the context is important.

[12] The output power is also referred to as the electrical power output and symbolized as P_{elec}.

Using this information, noting that the value for c_p is given, and realizing that the ΔT is 2°F allows the following substitution.

$$\dot{m} = \frac{\dot{Q}_{\text{cond}}}{c_p \Delta T} = \frac{(2333\ \text{MW})\left(\frac{10^6\ \text{W}}{1\,\text{MW}}\right)\left(\frac{3.413\,\text{Btu}}{\text{W-hr}}\right)}{\left(1.00\ \frac{\text{Btu}}{\text{lbm-}^\circ\text{F}}\right)(2^\circ\ \text{F})} = \boxed{3.98 \times 10^9\ \text{lbm/hr}}$$

Answer is B.

NUCLEAR POWER SYSTEMS–10

Certain *engineered safety features* of this plant require electric power to operate. They are designed such that they are able to operate on AC or DC power. *Probability risk assessment* (PRA) studies for this plant show a 10^{-3} probability of DC power (i.e., battery power) not being available at any time during plant operation and a 10^{-2} probability of a "failure to start on demand" by the AC diesel generators. Related reliability studies indicate the occurrence of a failure of off-site power once per calendar year.[13] Further, plant emergency systems are required once in every ten of these failures.

What is the probability this plant's emergency systems will not operate due to a lack of electric power in any given year?

(A) 0.075×10^{-5}
(B) 0.100×10^{-5}
(C) 1.000×10^{-5}
(D) 1.100×10^{-2}

Solution:

First, determine how often, during plant operation, the emergency systems will be "demanded."

[13] Off-site power is the normal power supply to emergency systems.

From the problem statement,

$$\text{demand rate} = \left(\frac{1\text{ off-site failure}}{1\text{ yr}}\right)\left(\frac{1\text{ demand}}{10\text{ off-site failures}}\right) = 0.1\text{ demands/yr}$$

The plant, however, is not operational all 365 days of the year. From the plant data, the availability is 75%.[14] This means the plant is exposed to electric power failure and the concomitant need for the back up power only 75% of the time. Thus, the above demand rate must be adjusted as follows.

$$\text{adjusted demand rate} = \left(0.1\frac{\text{demands}}{\text{yr}}\right)(0.75) = 0.075\text{ demands/yr}$$

The probability of emergency electric power failure (P_{fe}) occurs only if both of the AC and DC power backups fail. (Note the "or" condition in the problem statement.) Thus, from the given data,

$$P_{fe} = (10^{-3})(10^{-2}) = 10^{-5}\text{ failures/demand}$$

Note: If the emergency systems could operate on AC only or DC only, a failure of either would result in the non-operability of the system. In such a case, the probabilities would be summed, not multiplied.

Lastly, the probability of emergency system failure in a year (P_f) is the probability of emergency electric power failure multiplied by the adjusted demand rate.

$$\begin{aligned} P_f &= P_{fe}(\text{adjusted demand rate}) \\ &= \left(10^{-5}\frac{\text{failures}}{\text{demand}}\right)\left(0.075\frac{\text{demands}}{\text{yr}}\right) \\ &= 0.075\times10^{-5}\text{ failures/yr} \end{aligned}$$

The unit's failures per year, in this case, are equivalent to the probability of the plant's emergency systems not operating due to lack of electric power in any given year.[15]

[14] See the Nuclear Fuel Management section for information on *availability factors* (AF) and *capacity factors* (CF).

[15] "Demands" and "failures" are not units requiring canceling. They are used for clarification.

Therefore, the probability desired is

$$P = \boxed{0.075\times10^{-5}}$$

Answer is A.[16]

NUCLEAR POWER SYSTEMS–11

What is the *burnup rate* (BR) of a 500 MW_e U-235 fueled reactor with a recoverable energy per fission (E_r) of 200 MeV and a plant efficiency (η_{th}) of 25%?

(A) 2 g/day
(B) 200 g/day
(C) 525 g/day
(D) 2100 g/day

Solution:

First, calculate the *fission rate* (FR) for a given power—in this case for MW.

$$\begin{aligned} \text{FR} &= (P\text{ MW})\left(\frac{10^6\text{ J}}{1\text{ MW}\cdot\text{s}}\right)\left(\frac{1\text{ MeV}}{1.60\times10^{-13}\text{ J}}\right) \\ &\quad\times\left(\frac{1\text{ fission}}{200\text{ MeV}}\right)\left(\frac{86\,400\text{ s}}{1\text{ day}}\right) \\ &= \left((P)(2.70\times10^{21})\frac{\text{fissions}}{\text{day}}\right) \end{aligned}$$

In words, the power in megawatts multiplied by the factor in parenthesis gives the fissions/day required to produce that power.[17] The burnup rate is the fission rate adjusted to the appropriate units.

$$\begin{aligned} \text{BR} &= \left((P)(2.70\times10^{21})\frac{\text{fissions}}{\text{day}}\right)\left(\frac{1\text{ atom }^{235}\text{U}}{1\text{ fission}}\right) \\ &\quad\times\left(\frac{235\text{ g }^{235}\text{U}}{6.02\times10^{23}\text{ atoms }^{235}\text{U}}\right) \\ &= 1.05P\text{ g/day}\quad[P\text{ in MW}] \end{aligned}$$

[16] Some texts and studies define operational failures (OF) or other similar terms to account for the operational time of any given plant. The approach used here is one of unit analysis to avoid introducing unnecessary terms.

[17] The power referred to is the thermal power (P_{th}) since the reactor supplies all the energy generated, including that lost as heat. The subscript is often not displayed.

This formula, though approximate, provides adequate results for most calculations.[18]

Calculating the burnup rate for the information provided gives

$$\begin{aligned}\text{BR} &= 1.05P\,\frac{\text{g}}{\text{day}}\\ &= \left((1.05)\left(\frac{P_{\text{th}}}{\eta_{\text{th}}}\right)\frac{\text{g}}{\text{day}}\right)\\ &= \left((1.05)\left(\frac{500\ \text{MW}}{0.25}\right)\frac{\text{g}}{\text{day}}\right)\\ &= \boxed{2100\ \text{g/day}}\end{aligned}$$

Answer is D.

NUCLEAR POWER SYSTEMS–12

A steam generator is designed to support a steam mass flow rate ($\dot{m}$) of 122×10^6 lbm/hr with a maximum moisture content of 0.25%. The operating pressure is 1100 psia (P_{SG}) at a design feedwater inlet temperature (T_{FW}) of 440°F. What is the thermal power rating of the steam generator?

(A) 8.0×10^{10} Btu/hr
(B) 9.4×10^{10} Btu/hr
(C) 9.5×10^{10} Btu/hr
(D) 10.0×10^{10} Btu/hr

Solution:

The thermal power rating, or more commonly the heat transfer rate, is given by

$$\dot{Q} = \dot{m}\left(h_{\text{out}} - h_{\text{in}}\right)$$

The mass flow rate is given. The enthalpies need to be determined.

Interpolate between the values given in Table 1.1 in App. 1.A. The interpolating factor (F) for determining the 1100 psi entries is[19]

$$\begin{aligned}\text{F} &= \frac{1100\,\frac{\text{lbf}}{\text{in}^2} - 961.5\,\frac{\text{lbf}}{\text{in}^2}}{1131.8\,\frac{\text{lbf}}{\text{in}^2} - 961.5\,\frac{\text{lbf}}{\text{in}^2}}\\ &= \frac{138.5\,\frac{\text{lbf}}{\text{in}^2}}{170.3\,\frac{\text{lbf}}{\text{in}^2}}\\ &= 0.81\end{aligned}$$

Thus, for h_f,

$$\begin{aligned}h_{f,1100} &= (0.81)\left(562.0\,\frac{\text{Btu}}{\text{lbm}} - 536.4\,\frac{\text{Btu}}{\text{lbm}}\right) + 536.4\,\frac{\text{Btu}}{\text{lbm}}\\ &= 557.1\ \text{Btu/lbm}\end{aligned}$$

For h_g,

$$\begin{aligned}h_{g,1100} &= (0.81)\left(1187.0\,\frac{\text{Btu}}{\text{lbm}} - 1193.8\,\frac{\text{Btu}}{\text{lbm}}\right) + 1193.8\,\frac{\text{Btu}}{\text{lbm}}\\ &= 1188.3\ \text{Btu/lbm}\end{aligned}$$

For h_{fg},

$$\begin{aligned}h_{fg,1100} &= h_{g,1100} - h_{f,1100}\\ &= 1188.3\,\frac{\text{Btu}}{\text{lbm}} - 557.1\,\frac{\text{Btu}}{\text{lbm}}\\ &= 631.2\ \text{Btu/lbm}\end{aligned}$$

Using this line of 1100 psi data and the fact that the moisture content of 0.25% results in a quality x of

$$\begin{aligned}x &= 1 - \text{moisture content}\\ &= 1 - 0.0025\\ &= 0.9975\end{aligned}$$

[18] The actual *consumption rate* (CR) is higher since radiative capture occurs. To obtain CR from the burnup rate multiply BR by 1+ α, where $\alpha = \sigma_a / \sigma_f$. See the Nuclear Theory section for more information.

[19] As in all linear interpolation of the form $y = mx + b$ where "m" is the slope, the "direction" of the interpolation (e.g., from 961.5 to 1131.8 in this problem) is unimportant. Consistency, however, throughout the problem is important in order to obtain the correct result.

The enthalpy h_{out} can be calculated from the following.

$$\begin{aligned} h_{out,1100} &= h_{f,1100} + x h_{fg,1100} \\ &= 557.1\,\frac{\text{Btu}}{\text{lbm}} + (0.9975)\left(631.2\,\frac{\text{Btu}}{\text{lbm}}\right) \\ &= 1186.7\ \text{Btu/lbm} \end{aligned}$$

Note: The value calculated for $h_{out,1100}$ differs from h_g by less than 1.0%. This is because the quality of steam exiting a steam generator is designed to be high—moisture content low—to prevent/minimize water impingement damage to turbine blading. Thus, this type of interpolation using the quality of the steam from the exit of a steam generator can be ignored and the value of h_g associated with the appropriate pressure in the steam tables can be used directly. [20]

For h_{in}, use the value of h_f from Table 1.1 in App. 1.A correlating with 440°F (T_{FW}). Thus,

$$h_{in} = 419.0\ \text{Btu/lbm}$$

Note: The pressure associated with h_f is 381.2 psia while the actual pressure is 1100 psia. Tables exist for determining the enthalpy of such subcooled liquids, but errors introduced by associating the enthalpy of a saturated liquid at the given temperature with such a liquid at a higher pressure are inconsequential for most engineering calculations.

Therefore, substituting the given and calculated values gives

$$\begin{aligned} \dot{Q} &= \dot{m}(h_{out} - h_{in}) \\ &= \left(122\times10^{6}\,\frac{\text{lbm}}{\text{hr}}\right)\left(1188.3\,\frac{\text{Btu}}{\text{lbm}} - 419.0\,\frac{\text{Btu}}{\text{lbm}}\right) \\ &= \boxed{9.4\times10^{10}\ \text{Btu/hr}} \end{aligned}$$

Answer is B.

[20] The quality of steam on the outlet of a turbine is on the order of 25% and cannot be ignored in calculations. The method is identical to that used in this problem.

NUCLEAR POWER SYSTEMS–13

A 1200 MW$_e$ *pressurized water reactor* (PWR) with an overall thermal efficiency (η_{th}) of 30% is designed to supply saturated steam at a pressure of 1200 psia to two steam turbines. The exhaust of each turbine is cooled to near saturated liquid conditions at 4 psia in a single condenser. Assuming this PWR operates on an ideal Rankine cycle, calculate the thermal power transferred by the condenser.

(A) 9.05×10^{3} Btu/hr
(B) 2.65×10^{9} Btu/hr
(C) 9.05×10^{9} Btu/hr
(D) 2.65×10^{15} Btu/hr

Solution:

First, consider the ideal Rankine cycle shown.

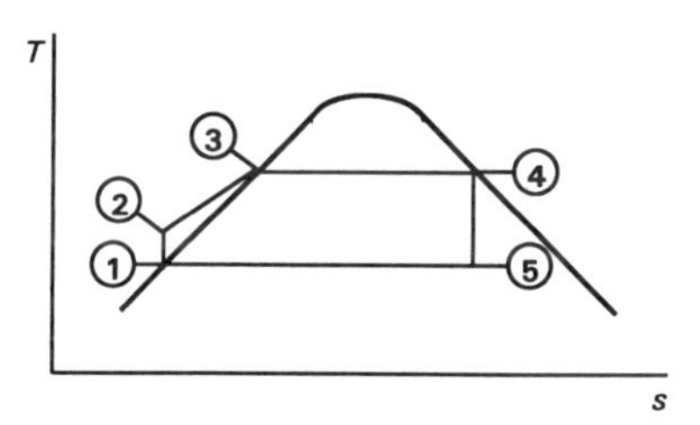

The numbered points on the Rankine cycle drawing correlate with the points indicated on the *secondary system* drawing shown.

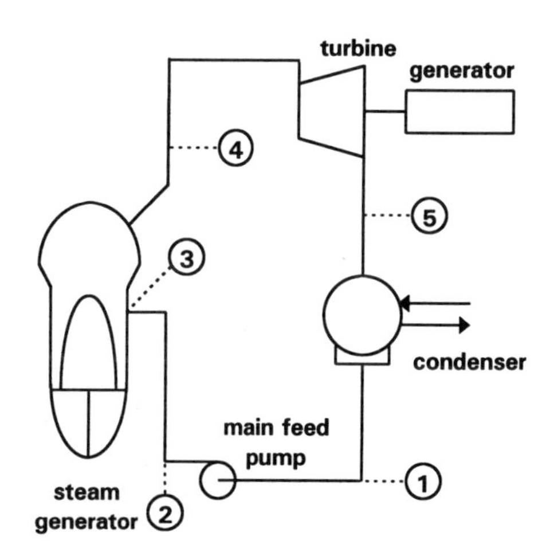

The heat transfer rate, or thermal power, of the condenser is given by

$$\dot{Q}_{\text{cond}} = \dot{m}(h_5 - h_1)$$

The value of h_1 is tabulated, h_5 and $\dot{m}$ must be determined through calculation.

From Table 1.2 in App. 1.A, the value for the enthalpy of a saturated liquid (h_f) at 4 psia, equivalent to h_1, is

$$h_1 = 120.9 \text{ Btu/lbm}$$

In order to determine h_5, additional information is required, specifically the quality. Therefore, note that[21]

$$s_4 = s_5$$

Using Table 1.2 in App. 1.A to determine s_4, the entropy of a saturated vapor at 1200 psia, gives

$$s_4 = 1.3673 \text{ Btu/lbm-}^\circ\text{R}$$

$$s_5 = 1.3673 \text{ Btu/lbm-}^\circ\text{R}$$

To determine the quality x at 4 psia, use data from Table 1.2 in App. 1.A and the following.

$$\begin{aligned} s_5 &= s_{5,f} + x s_{5,fg} \\ x &= \frac{s_5 - s_{5,f}}{s_{5,fg}} \\ &= \frac{1.3673 \frac{\text{Btu}}{\text{lbm-}^\circ\text{R}} - 0.2198 \frac{\text{Btu}}{\text{lbm-}^\circ\text{R}}}{1.6426 \frac{\text{Btu}}{\text{lbm-}^\circ\text{R}}} \\ &= 0.70 \end{aligned}$$

Use this quality to determine the value of h_5.

$$\begin{aligned} h_5 &= h_{5,f} + x h_{5,fg} \\ &= 120.9 \frac{\text{Btu}}{\text{lbm}} + (0.7)\left(1006.4 \frac{\text{Btu}}{\text{lbm}}\right) \\ &= 825.4 \text{ Btu/lbm} \end{aligned}$$

[21] For an ideal simple Rankine cycle, the process through the steam turbine is adiabatic and reversible, thus $\Delta s = 0$.

To determine the mass flow rate, use the portion of the process from point 4 to point 1 and

$$\dot{Q}_r = \dot{m}(h_4 - h_1)$$

$\dot{Q}_r$ = reactor heat input[22]

The reactor heat input is given by

$$\begin{aligned} \dot{Q}_r &= \frac{P_{\text{out}}}{\eta_{\text{th}}} \\ &= \frac{1200 \text{ MW}}{0.30} \\ &= 4000 \text{ MW} \end{aligned}$$

The enthalpy at point 4 is the enthalpy of a saturated vapor at 1200 psia. From Table 1.2 in App. 1.A

$$h_4 = 1183.9 \text{ Btu/lbm}$$

Rearranging the reactor heat input formula to solve for the mass flow rate gives

$$\begin{aligned} \dot{Q}_r &= \dot{m}(h_4 - h_1) \\ \dot{m} &= \frac{\dot{Q}_r}{h_4 - h_1} \\ &= \frac{(4000 \text{ MW})\left(\frac{10^6 \text{ W}}{1 \text{ MW}}\right)\left(\frac{3.413 \frac{\text{Btu}}{\text{hr}}}{1 \text{ W}}\right)}{1183.9 \frac{\text{Btu}}{\text{lbm}} - 120.9 \frac{\text{Btu}}{\text{lbm}}} \\ &= 12.84 \times 10^6 \text{ lbm/hr} \end{aligned}$$

Substituting the gathered information gives

$$\begin{aligned} \dot{Q}_{\text{cond}} &= \dot{m}(h_5 - h_1) \\ &= \left(12.84 \times 10^6 \frac{\text{lbm}}{\text{hr}}\right)\left(825.4 \frac{\text{Btu}}{\text{lbm}} - 120.9 \frac{\text{Btu}}{\text{lbm}}\right) \\ &= \boxed{9.05 \times 10^9 \text{ Btu/hr}} \end{aligned}$$

Answer is C.

[22] This includes the energy used by the turbine (P_{out} or P_{elec}) and the heat rejected by the condenser.

Nuclear Power Systems (NPS) problems 14–16 are based on the following information and illustration.

The rate of fission in a given fuel rod, and hence the rate of heat production, varies between fuel rods and is also a function of the position of the fuel rod in the core as given by the following expression.[23]

$$q'''(\vec{r}) = E_d \int_0^\infty \Sigma_{f,r}(E)\varphi(\vec{r},E)dE$$

q''' = power density, or power production per unit volume, as a function of position in the core[24]
E_d = energy deposited locally in the fuel per fission and normally considered to be ≈ 180 MeV
$\Sigma_{f,r}$ = macroscopic fission cross section of the fuel rod as a function of energy
φ = energy dependent flux as a function of radial position

Since, in a thermal reactor, most fissions are produced by thermal neutrons; and assuming a bare, finite cylinder as the reactor shape; and finally, calculating the total average macroscopic fission cross section for the core ($\Sigma_{f,25}$) to account for the rods ($\Sigma_{f,r}$) as well as all other components as if they were a homogeneous mixture; results in the following simplified expression.[25]

$$q'''(r,z) = \left(\frac{1.16\,P\,E_d}{H\,a^2\,n\,E_r}\right) J_0\left(\frac{2.405r}{R}\right)\cos\left(\frac{\pi z}{H}\right)$$

P = total power
H = height of core [fuel rod height]
a = radius of the fuel rods
n = number of fuel rods
E_r = energy recoverable per fission ≈ 200 MeV
J_0 = Bessel function of the first kind of order zero
r = radial distance from the axis of the core
R = radius of the core
z = axial position along a given fuel rod usually represented as shown[26]

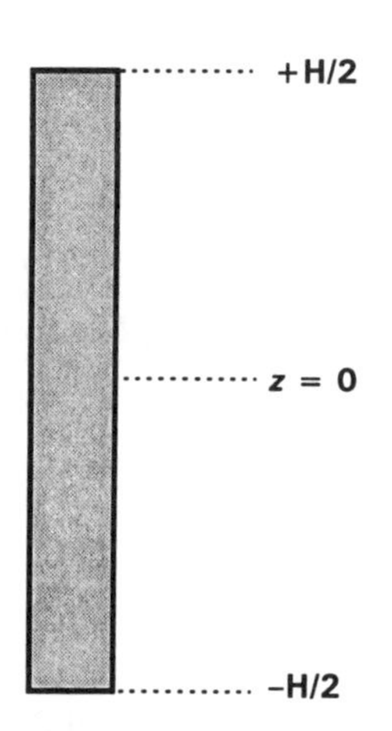

NUCLEAR POWER SYSTEMS–14

A 1000 MW_{th} research reactor is of cylindrical design, 6 ft in height and 8 ft in diameter. It contains 200 fuel assemblies each with 150 UO_2 cylindrical fuel rods with a diameter of 0.4 in. Assume the assemblies are uniformly distributed throughout the core and no *blanket* or *reflector* exists.[27] What is the power density in a fuel rod on the axis of the core, three-quarters of the distance from the bottom of the rod?

(A) 1.27×10^7 Btu/hr-ft^3
(B) 3.57×10^7 Btu/hr-ft^3
(C) 5.04×10^7 Btu/hr-ft^3
(D) 1.48×10^8 Btu/hr-ft^3

Solution:

The applicable formula is

$$q'''(r,z) = \left(\frac{1.16\,P\,E_d}{H\,a^2\,n\,E_r}\right) J_0\left(\frac{2.405r}{R}\right)\cos\left(\frac{\pi z}{H}\right)$$

The following items are given information: P, E_d, H, a, and E_r.

The radius of the core is calculated from the diameter (D).

$$\begin{aligned} R &= \frac{D}{2} \\ &= \frac{8 \text{ ft}}{2} \\ &= 4 \text{ ft} \end{aligned}$$

[23] The fuel rod is also referred to as a fuel pin, or simply pin. Rod when used alone may refer to the neutron absorber used to control the fission chain reaction. See the Nuclear Theory section for more information.
[24] The units depend on the desired units in the result; although, in most heat transfer calculations the English Engineering System continues to be used, hence Btu/hr-ft³.
[25] This cylindrical shape is used as the base assumption for many initial heat removal calculations.
[26] Setting the zero point mid-axis eases calculations as will be shown in the associated problems.
[27] In other words, this is a bare reactor.

The number of fuel rods is calculated as follows.

$$\begin{aligned} n &= (\text{no. of assemblies})\left(\frac{\text{fuel rods}}{\text{assembly}}\right) \\ &= (200\ \text{assemblies})\left(150\ \frac{\text{fuel rods}}{\text{assembly}}\right) \\ &= 30\,000\ \text{fuel rods} \end{aligned}$$

The rod in question is on the axis (i.e., $r = 0$) giving for the second term in the equation

$$\begin{aligned} \text{J}_0\left(\frac{2.405r}{R}\right) &= \text{J}_0\left(\frac{(2.405)(0\ \text{ft})}{4\ \text{ft}}\right) \\ &= \text{J}_0(0) \\ &= 1 \end{aligned}$$

The value for J_0 (0) was obtained from Table 1.3 in App. 1.A.

For the cosine term, the position referred to is three-quarters from the bottom of the fuel rod.

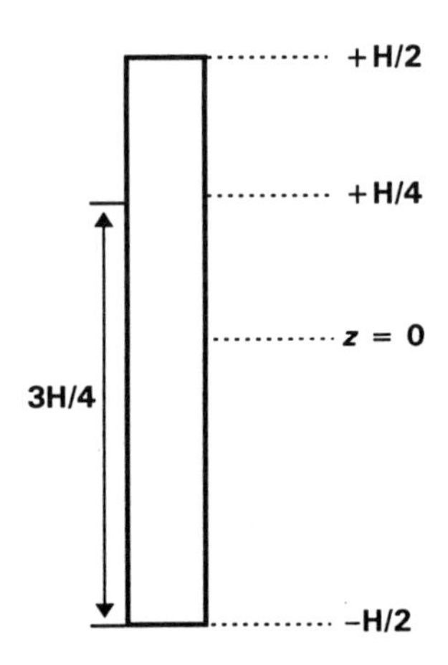

Thus, $z = H/4$ and

$$\begin{aligned} \cos\left(\frac{\pi z}{H}\right) &= \cos\left(\frac{(\pi)\left(\frac{H}{4}\right)}{H}\right) \\ &= \cos\left(\frac{\pi}{4}\right) \\ &= 0.707 \end{aligned}$$

Substituting gives

$$q'''(r,z) = \left(\frac{1.16PE_d}{Ha^2nE_r}\right)\text{J}_0\left(\frac{2.405r}{R}\right)\cos\left(\frac{\pi z}{H}\right)$$

$$\begin{aligned} q'''(0, H/4) &= \left(\frac{(1.16)(1000\ \text{MW})(180\ \text{MeV})}{(6\ \text{ft})\left(\left(\frac{0.4\ \text{in}}{2}\right)\left(\frac{1\ \text{ft}}{12\ \text{in}}\right)\right)^2(30\,000)(200\ \text{MeV})}\right) \\ &\quad \times (1)(0.707) \\ &= 14.8\ \text{MW/ft}^3 \\ &= \left(14.8\ \frac{\text{MW}}{\text{ft}^3}\right)\left(\frac{10^6\ \text{W}}{1\ \text{MW}}\right)\left(\frac{3.413\ \frac{\text{Btu}}{\text{hr}}}{\text{W}}\right) \\ &= \boxed{5.04\times 10^7\ \text{Btu/hr-ft}^3} \end{aligned}$$

Answer is C.

NUCLEAR POWER SYSTEMS–15

What is the power density at full power of a fuel rod located 2 ft from the axis, of the core described in Prob. 14, at the point of its maximum power density?

(A) 1.2×10^7 Btu/hr-ft^3
(B) 4.8×10^7 Btu/hr-ft^3
(C) 5.8×10^7 Btu/hr-ft^3
(D) 1.4×10^8 Btu/hr-ft^3

Solution:

Again the applicable equation is

$$q'''(r,z) = \left(\frac{1.16PE_d}{Ha^2nE_r}\right)\text{J}_0\left(\frac{2.405r}{R}\right)\cos\left(\frac{\pi z}{H}\right)$$

The values of items in the first term of the equation remain the same as in Prob. 14. The second term must be solved for a fuel rod 2 ft from the axis.

Thus,

$$J_0\left(\frac{2.405r}{R}\right) = J_0\left(\frac{(2.405)(2\text{ ft})}{4\text{ ft}}\right) = J_0(1.202) = 0.6710$$

The value for $J_0(1.202)$ was obtained by interpolation from Table 1.3 in App. 1.A.

Consider the third term. The point of maximum power density for a given set of conditions occurs when this cosine term is maximum (i.e., equal to one). Thus,

$$\cos\left(\frac{\pi z}{H}\right) = 1$$

$$\frac{\pi z}{H} = \cos^{-1}(1) = 0$$

$$z = \frac{H(0)}{\pi} = \frac{(6\text{ ft})(0)}{\pi} = 0$$

Substituting gives

$$q'''(r,z) = \left(\frac{1.16PE_d}{Ha^2nE_r}\right)J_0\left(\frac{2.405r}{R}\right)\cos\left(\frac{\pi z}{H}\right)$$

$$q'''(2,0) = \left(\frac{(1.16)(1000\text{ MW})(180\text{ MeV})}{(6\text{ ft})\left(\left(\frac{0.4\text{ in}}{2}\right)\left(\frac{1\text{ ft}}{12\text{ in}}\right)\right)^2(30\,000)(200\text{ MeV})}\right) \times (0.6710)(1)$$

$$= 14.0\,\frac{\text{MW}}{\text{ft}^3} = \left(14.0\,\frac{\text{MW}}{\text{ft}^3}\right)\left(\frac{10^6\text{ W}}{1\text{ MW}}\right)\left(\frac{3.413\,\frac{\text{Btu}}{\text{hr}}}{1\text{ W}}\right)$$

$$= \boxed{4.78\times10^7\text{ Btu/hr-ft}^3 \quad (4.8\times10^7\text{ Btu/hr-ft}^3)}$$

Answer is B.

NUCLEAR POWER SYSTEMS–16

What is the total power of the fuel rod located on the axis of the core when the core is operated at full power?

(A) 69×10^3 Btu/hr
(B) 85×10^3 Btu/hr
(C) 170×10^3 Btu/hr
(D) 240×10^3 Btu/hr

Solution:

Recall that the power density is given by

$$q'''(r,z) = \left(\frac{1.16PE_d}{Ha^2nE_r}\right)J_0\left(\frac{2.405r}{R}\right)\cos\left(\frac{\pi z}{H}\right)$$

The first two terms are constant at a given power level P for a given fuel rod position r in the core. Let these first two terms be defined as constant C, for convenience. Thus,

$$q'''(r,z) = \text{C}\cos\left(\frac{\pi z}{H}\right)$$

Noting that the total power for any given fuel rod is the power density integrated over the length of the fuel rod multiplied by the cross sectional area (A) gives

$$\dot{Q}_r(r) = A\int_{-H/2}^{H/2} q'''(r,z)\,dz$$

Since the radius of a fuel rod is given by a, the cross sectional area is

$$A = \pi a^2$$

Using this information, the defined constant C, and substituting gives

$$\dot{Q}_r(r) = A\int_{-H/2}^{H/2} q'''(r,z)\,dz$$

$$= \pi a^2 \int_{-H/2}^{H/2} \mathrm{C}\cos\left(\frac{\pi z}{H}\right)dz = \pi a^2\,\mathrm{C}\int_{-H/2}^{H/2}\cos\left(\frac{\pi z}{H}\right)dz$$

$$= \pi a^2\mathrm{C}\left(\left(\frac{H}{\pi}\right)\sin\left(\frac{\pi z}{H}\right)\Big|_{-H/2}^{H/2}\right)$$

$$= \pi a^2\mathrm{C}\left(\left(\frac{H}{\pi}\right)(1-(-1))\right)$$

$$= \pi a^2\mathrm{C}\left(\frac{2H}{\pi}\right)$$

Expand the constant C and cancel terms.

$$\dot{Q}_r(r) = \pi a^2\left(\left(\frac{1.16\,PE_d}{Ha^2nE_r}\right)\mathrm{J}_0\left(\frac{2.405r}{R}\right)\right)\left(\frac{2H}{\pi}\right)$$

$$= \left(\frac{2.32PE_d}{nE_r}\right)\mathrm{J}_0\left(\frac{2.405r}{R}\right)$$

The question concerns the rod on the axis of the core (i.e., at $r = 0$) where $\mathrm{J}_0(0) = 1$ (from Table 1.3 in App. 1A). The result is

$$\dot{Q}_r(r) = \left(\frac{2.32\,PE_d}{nE_r}\right)\mathrm{J}_0\left(\frac{2.405r}{R}\right)$$

$$= \left(\frac{(2.32)(1000\text{ MW})(180\text{ MeV})}{(30\,000)(200\text{ MeV})}\right)\mathrm{J}_0\left(\frac{(2.405)(0\text{ ft})}{4\text{ ft}}\right)$$

$$= 69.6\times10^{-3}\text{ MW}$$

$$= (69.6\times10^{-3}\text{ MW})\left(\frac{10^6\text{ W}}{1\text{MW}}\right)\left(\frac{3.413\frac{\text{Btu}}{\text{hr}}}{1\text{ W}}\right)$$

$$= \boxed{237.5\times10^3\text{ Btu/hr}\quad(240\times10^3\text{ Btu/hr})}$$

Answer is D.

Note: The amount of energy production in a core depends on, among other things, the neutron flux distribution as shown in NPS problems 14, 15, and 16. Since radioactive decay energy (decay heat production) depends on the quantity of decay products in a given region, which in turn depends on the energy production in this same region, it therefore also depends on the neutron flux. Thus, problems where a determination of decay heat production is required follow the same general format and methods used in these problems.

Nuclear Power Systems (NPS) problems 17 and 18 are based on the following information and illustrations.

A plate-type fuel element is shown from two views with the dimensions as indicated.

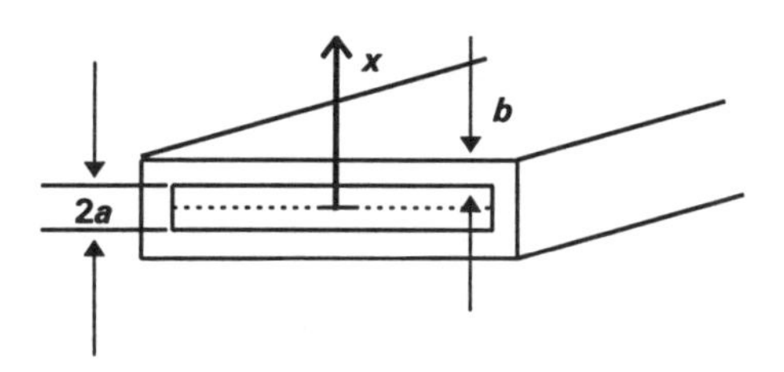

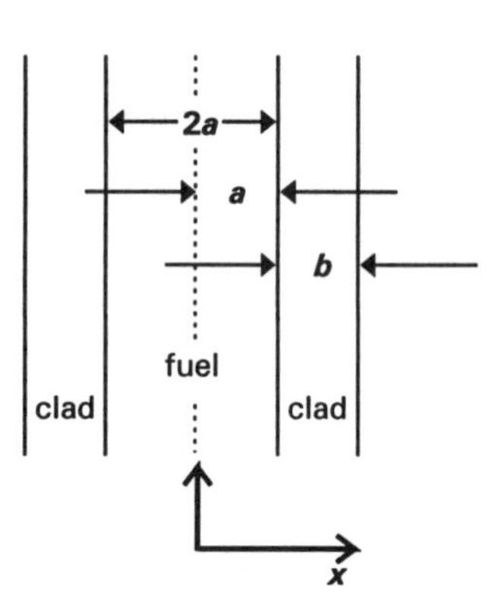

Plant Data

$2a = 0.40$ in

$b = 0.002$ ft

A = heat transfer area of one face of the fuel plate = 4 ft^2

Scientific Data

thermal conductivity (k):

Zircaloy-4 $k_c = 5$ Btu/hr-ft-°F

UO_2 fuel $k_f = 3$ Btu/hr-ft-°F

film coefficient (h):

water $h_{H2O} = 2000$ Btu/hr-ft^2-°F

NUCLEAR POWER SYSTEMS–17

What is the total thermal resistance R from the center of the fuel to the coolant? (Ignore any resistance due to the bonding between the fuel and the cladding, which is normally very small.)

(A) 7.94×10^{-4} hr-°F/Btu
(B) 9.19×10^{-4} hr-°F/Btu
(C) 2.78×10^{-3} hr-°F/Btu
(D) 3.18×10^{-3} hr-°F/Btu

Solution:

The total thermal resistance is defined as follows.

$$R = \begin{pmatrix}\text{thermal resistance}\\ \text{of the fuel meat}\\ \text{half-thickness}\end{pmatrix} + \begin{pmatrix}\text{thermal resistance}\\ \text{the cladding}\end{pmatrix} + \begin{pmatrix}\text{thermal resistance}\\ \text{for convective}\\ \text{heat transfer}\end{pmatrix}$$

In terms of the variables given,

$$R = \frac{a}{2k_f A} + \frac{b}{k_c A} + \frac{1}{hA}$$

Note: The one-half in the term for the thermal resistance of the fuel meat is due to the symmetry of the plate-type construction (i.e., no heat flows at the centerline of the fuel) or in mathematical terms, one of the boundary conditions is that

$$\frac{dT}{dx} = 0 \quad \text{at} \quad x = 0$$

T = temperature

Substituting the given values gives

$$R = \frac{a}{2k_f A} + \frac{b}{k_c A} + \frac{1}{hA}$$

$$= \frac{\left(\frac{0.40 \text{ in}}{2}\right)\left(\frac{1 \text{ ft}}{12 \text{ in}}\right)}{(2)\left(3\frac{\text{Btu}}{\text{hr-ft-°F}}\right)(4 \text{ ft}^2)} + \frac{0.002 \text{ ft}}{\left(5\frac{\text{Btu}}{\text{hr-ft-°F}}\right)(4 \text{ ft}^2)} + \frac{1}{\left(2000\frac{\text{Btu}}{\text{hr-ft}^2\text{-°F}}\right)(4 \text{ ft}^2)}$$

$$= 6.94 \times 10^{-4} \frac{\text{hr-°F}}{\text{Btu}} + 1.00 \times 10^{-4} \frac{\text{hr-°F}}{\text{Btu}} + 1.25 \times 10^{-4} \frac{\text{hr-°F}}{\text{Btu}}$$

$$= \boxed{9.19 \times 10^{-4} \text{ hr-°F/Btu}}$$

Answer is B.

NUCLEAR POWER SYSTEMS–18

The temperature profile of this plate-type fuel element is as follows.

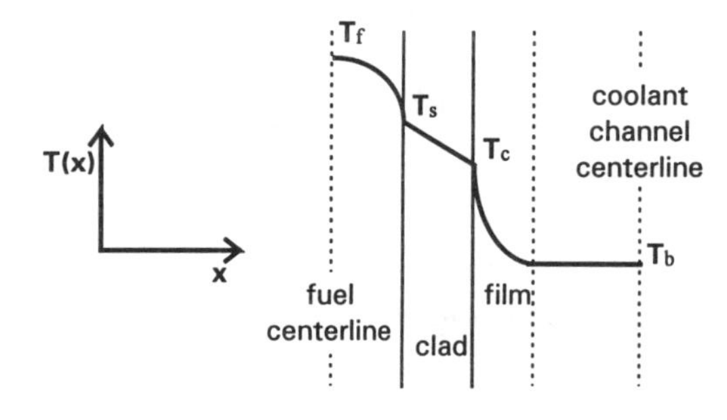

The ΔT between the bulk temperature (T_b) and the fuel temperature (T_f) at the centerline cannot exceed 100°F during steady state conditions, at the design T_b, in order to maintain the integrity of the fuel assembly.[28]

[28] That is, to keep the fuel inside the cladding. At the extreme, what this means is that to avoid melting the fuel and its surrounding matrix support structure the temperature limits must be observed.

What is the limiting heat transfer rate for a single fuel plate?

(A) 32 kW
(B) 40 kW
(C) 48 kW
(D) 64 kW

Solution:

The total heat transfer is given by[29]

$$\dot{Q} = \frac{T_f - T_b}{R} = \frac{T_f - T_b}{\frac{a}{2k_f A} + \frac{b}{k_c A} + \frac{1}{hA}} = \Delta T_{fb}/R$$

The total thermal resistance R was calculated in the Prob. 17 and the temperature difference is given. Substituting,

$$\dot{Q} = \frac{\Delta T_{fb}}{R} = \frac{100°\text{F}}{9.19\times10^{-4}\,\frac{\text{hr-°F}}{\text{Btu}}} = 108.8\times10^{3}\,\frac{\text{Btu}}{\text{hr}}$$

$$= \left(108.8\times10^{3}\,\frac{\text{Btu}}{\text{hr}}\right)\left(\frac{1\,\text{W}}{3.413\,\frac{\text{Btu}}{\text{hr}}}\right)\left(\frac{1\,\text{kW}}{10^{3}\,\text{W}}\right) = 31.88\,\text{kW}$$

This is the heat transfer rate, or power, from one-half of the fuel plate; therefore, for the entire fuel plate,

$$\dot{Q}_{\text{total}} = \dot{Q}\times2 = (31.88\,\text{kW})(2) = \boxed{63.78\,\text{kW}\ \ (64\,\text{kW})}$$

Answer is D.

[29] The formula is analogous to I = V/R in electrical theory. See the theory texts in the References section.

NUCLEAR POWER SYSTEMS–19

A simplified system *fault tree* leading to a pipe rupture is shown below.

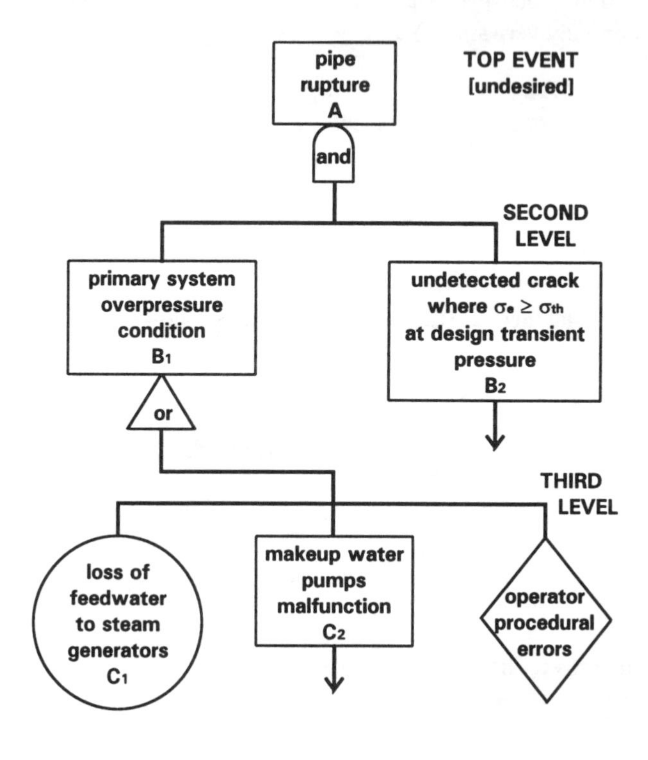

Symbology
circle = the probability of occurrence is known or can be calculated
rectangle = the probability must be traced from lower levels in the fault tree
diamond = the probability is not developed due to lack of information or the insignificance of the item
A, B, C = probability of the indicated event
σ_{th} = stress at the instant of fracture
σ_e = stress at the end of a crack

Using Boolean algebra, what is the probability of fault A?

(A) $(C_1 + C_2)B_2$
(B) $B_1 + B_2$
(C) $C_1 \cdot C_2 \cdot B_2$
(D) $C_1 + C_2 + B_2$

Solution:

Commence at the level of the event of interest, in this case the top event or pipe rupture. The probability of this event has been defined as A. This is the desired quantity.[30]

Consider the next level and the events in it that lead to the pipe rupture. Note that both events in level two must occur for the rupture to take place, as indicated by the "and" connection. The probability is then

$$A = B_1 \cdot B_2$$

In the next level, labeled the "third level," any one of the events occurring, as indicated by the "or" connection, means that the over-pressure condition will occur. Thus, the probability for the over-pressure condition is

$$B_1 = C_1 + C_2$$

Procedural errors by the operator have no associated probability in this case, likely due to the difficulty in quantifying this information, and are thus not included. Substituting this result into the first gives

$$\boxed{A = (C_1 + C_2)B_2}$$

Answer is A.

NUCLEAR POWER SYSTEMS–20

Consider a pipe rupture as an initiating event for a *loss-of-coolant accident* (LOCA). A reduced *event tree* showing the probabilities of the most likely consequences is shown. The probability of failure at each stage is labeled as P_1, P_2...with the probability of success (i.e., the system is available and functioning normally) represented by $1-P_2$, $1-P_3$, and so on. Radioactive release events are labeled I, II, III, and IV.

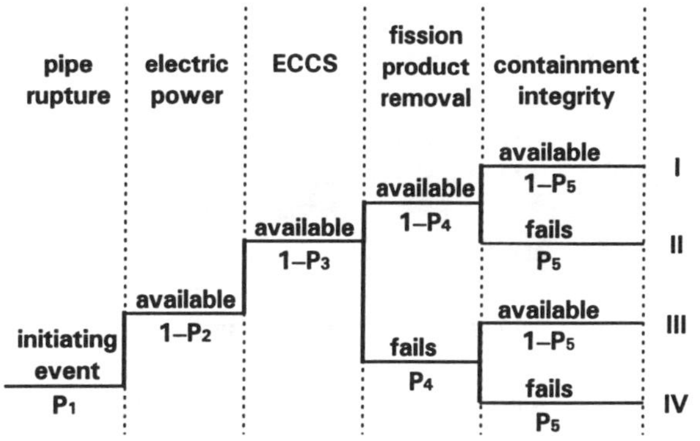

pipe rupture
electric power
ECCS
fission product removal
containment integrity
initiating event
P_1
available
$1-P_2$
available
$1-P_3$
available
$1-P_4$
fails
P_4
available
$1-P_5$
fails
P_5
available
$1-P_5$
fails
P_5
I
II
III
IV

Terminology
ECCS = emergency core-cooling system

What is the probability of the most severe event (i.e., radioactive release I, II, III, or IV) shown?

(A) P_1
(B) $(P_1)(P_5)$
(C) $(P_1)(1-P_2)(1-P_3)(P_4)(1-P_5)$
(D) $(P_1)(P_4)(P_5)$

Solution:

The overall probability of a chain of events, with the events labeled as I–IV, is the product of the probabilities in the individual chain. For example, the probability of event IV is

$$P_{IV} = (P_1)(1-P_2)(1-P_3)(P_4)(P_5)$$

Note that $(1-P_2)$ and $(1-P_3)$ represent the probability that failure does not occur.

Reductions of the event equations, such as P_{IV}, can occur due to the probabilities being very small—on the order of 10^{-4} and smaller. Therefore, $(1-P_2)$ and $(1-P_3)$ can be taken to be unity (i.e., one).[31] Thus, the probabilities of the events are

event	probability
I	P_1
II	$(P_1)(P_5)$
III	$(P_1)(P_4)$
IV	$(P_1)(P_4)(P_5)$

[30] An underlying assumption here is that all the events shown are mutually exclusive. See the statistics text in the References section for additional information.

[31] The effect of this assumption, assuming all the above probabilities to be 10^{-4} (a fairly large value in probability risk assessment, PRA, studies of this type) is an error in the final value of event IV of 0.02%.

The probabilities of each event are now known. What remains is to determine the most severe event. Severity can be defined in numerous ways, in this case, the event with the greatest release of radioactivity to the surrounding environment—and thus the greatest exposure to the public—is the most severe.

Engineered safety features, such as the emergency core-cooling system (ECCS), the fission product removal system, and the containment structure itself, minimize this severity when the reactor protection system alone cannot accommodate the transient or accident. Therefore, the event which includes the failure of the largest number of these safety features is the most severe.

The fission product removal system and the containment structure both fail in event IV; thus, the most severe event is event IV. From the table, the probability of event IV is

$$\boxed{(P_1)(P_4)(P_5)}$$

Answer is D .

NUCLEAR POWER SYSTEMS–21

A pressurized water reactor (PWR) has a primary coolant inventory of 300 000 kg at 15.5 MPa_a and 315°C. the containment structure's free volume is 55 000 m^3. What is the pressure load on the containment building following a loss-of-coolant accident (LOCA)?

Consider this calculation a scoping calculation to determine an approximate value prior to an in-depth design calculation. Therefore, the thermal energy absorbed by plant components, the containment structure itself, and any injected water as well as heat input by fission product decay are to be ignored.

(A) 100 kPa_a
(B) 200 kPa_a
(C) 400 kPa_a
(D) 500 kPa_a

Solution:

First determine the amount of steam produced if the final pressure is assumed to be 50 psig.[32]

[32] This pressure, 50 psig or 345 kPa_g, is representative of the design pressure for PWR cylindrical containment structures. Some spherical containment structures may be designed as high as approximately 72.5 psig or 500 kPa_g.

As the solution and most of the given data are in SI units, convert the 50 psig.[33]

$$50\,\frac{\text{lbf}}{\text{in}^2} = \left(50\,\frac{\text{lbf}}{\text{in}^2}\right)\left(\frac{4.4482\ \text{N}}{1\ \text{lbf}}\right)\left(\frac{1\ \text{in}}{2.54\ \text{cm}}\right)^2\left(\frac{100\ \text{cm}}{1\ \text{m}}\right)^2$$

$$= \left(3.45\times 10^5\,\frac{\text{N}}{\text{m}^2}\right)\left(\frac{1\ \text{Pa}}{1\,\frac{\text{N}}{\text{m}^2}}\right)\left(\frac{1\ \text{kPa}}{10^3\ \text{Pa}}\right) = 345\ \text{kPa}$$

From the SI unit steam Table 1.4 in App. 1.B, the specific enthalpy of the subcooled primary coolant at 315°C prior to the LOCA is determined by interpolation to be

$$h_{f,315} = \left(\frac{315^\circ\text{C} - 310^\circ\text{C}}{320^\circ\text{C} - 310\ ^\circ\text{C}}\right)\left(1461.5\,\frac{\text{kJ}}{\text{kg}} - 1344.0\,\frac{\text{kJ}}{\text{kg}}\right) + 1344.0\,\frac{\text{kJ}}{\text{kg}}$$
$$= 1402.8\ \text{kJ/kg}$$

At the assumed pressure of 345 kPa, the specific enthalpy of the saturated liquid from Table 1.5 in App. 1.B is

$$h_{f,345} = 582.0\ \text{kJ/kg}$$

The latent heat of vaporization is

$$h_{fg,345} = 2149.6\ \text{kJ/kg}$$

Calculate the steam produced during the LOCA.

$$m_{\text{steam}} = \frac{\left(\begin{array}{c}\text{mass of}\\ \text{primary fluid}\end{array}\right)\left(\begin{array}{c}\text{energy available}\\ \text{to produce steam}\end{array}\right)}{\begin{array}{c}\text{latent heat of vaporization}\\ \text{for the assumed pressure}\end{array}} = \frac{(m_{pri})(h_{f,pp} - h_{f,\text{LOCA}})}{h_{fg}}$$

$h_{f,pp}$ = the enthalpy of the subcooled primary coolant at primary system pressure
$h_{f,\text{LOCA}}$ = the enthalpy of the saturated liquid at the assumed LOCA containment pressure

[33] There are conversion factors directly from psia to kPa, although knowing a few basic conversions, such as those shown, is helpful when references are not available.

Substituting gives

$$m_{\text{steam}} = \frac{(m_{pri})(h_{f,pp} - h_{f,\text{LOCA}})}{h_{fg}} = \frac{(300\,000\text{ kg})\left(1402.8\,\frac{\text{kJ}}{\text{kg}} - 582.0\,\frac{\text{kJ}}{\text{kg}}\right)}{2149.6\,\frac{\text{kJ}}{\text{kg}}} = 114\,551\text{ kg}$$

The corresponding specific volume (v) is

$$\upsilon = \frac{\text{containment structure volume}}{\text{mass of steam}} = \frac{55\,000\text{ m}^3}{114\,551\text{ kg}} = 0.4801\text{ m}^3/\text{kg}$$

Using this specific volume and interpolating using the values from Table 1.5 in App. 1.B, the corresponding pressure is

$$P = \left(\frac{(414\text{ kPa} - 379\text{ kPa})}{0.4489\,\frac{\text{m}^3}{\text{kg}} - 0.4884\,\frac{\text{m}^3}{\text{kg}}}\right)\left(0.4801\,\frac{\text{m}^3}{\text{kg}} - 0.4884\,\frac{\text{m}^3}{\text{kg}}\right) + 379\text{ kPa} = 386.4\text{ kPa}$$

Repeat the process using this new pressure and its associated properties. The following table indicates the values obtained from Table 1.5 in App. 1.B following interpolation.

pressure (kPa)	υ_g (m³/kg)	h_f (kJ/kg)	h_{fg} (kJ/kg)	h_g (kJ/kg)
386.4	0.4800	599.1	2137.7	2736.8

Substituting to determine the mass of the steam gives[34]

$$m_{\text{stm}} = \frac{(m_{pri})(h_{f,pp} - h_{f,\text{LOCA}})}{h_{fg}} = \frac{(300\,000\text{ kg})\left(1402.8\,\frac{\text{kJ}}{\text{kg}} - 599.1\,\frac{\text{kJ}}{\text{kg}}\right)}{2137.7\,\frac{\text{kJ}}{\text{kg}}} = 112\,789\text{ kg}$$

Which results in a specific volume of

$$\upsilon = \frac{\text{containment structure volume}}{\text{mass of steam}} = \frac{55\,000\text{ m}^3}{112\,789\text{ kg}} = 0.4876\text{ m}^3/\text{kg}$$

This second-trial value for the specific volume is relatively unchanged from the first-trial value of 0.4801 m³/kg which correlated to a pressure of 386.4 kPa.[35] Thus the pressure added to the containment atmosphere by the steam is approximately 387 kPa_g (about 4 atm).

The total pressure is the sum of the partial pressures, in this case the air and steam pressures. Assuming an initial pressure of one atmosphere (101.3 kPa) in the containment building prior to the LOCA, the total pressure is

$$P_{\text{total}} \cong P_{\text{stm}} + P_{\text{atm}} \cong 387\text{ kPa}_g + 101\text{ kPa}_g \cong \boxed{488\text{ kPa}_a \quad (500\text{ kPa}_a)}$$

This is approximately 70 psia or 5 atm.[36]

Answer is D.

---

[34] The enthalpy of the subcooled liquid ($h_{f,pp}$) does not change as this is determined by the conditions in the primary system.

[35] Any number of trials can be completed depending on the accuracy desired. As this is meant to be a scoping calculation, the 0.0025 difference in the specific volumes provides adequate accuracy—within a few kPa's.

[36] The thermal energy considerations ignored in the problem are likely to reduce the pressure to approximately 300 kPa or 3 atm as would be evident in a more detailed calculation. Further, a containment designed for a pressure such as this, on the order of 500 kPa_a, would likely be spherical vice cylindrical.

Nuclear Power Systems (NPS) problems 22–24 are based on the following information and illustration.

The following stress-strain curve is typical of that for common stainless steels (adjusted for normal power plant operating temperatures).[37]

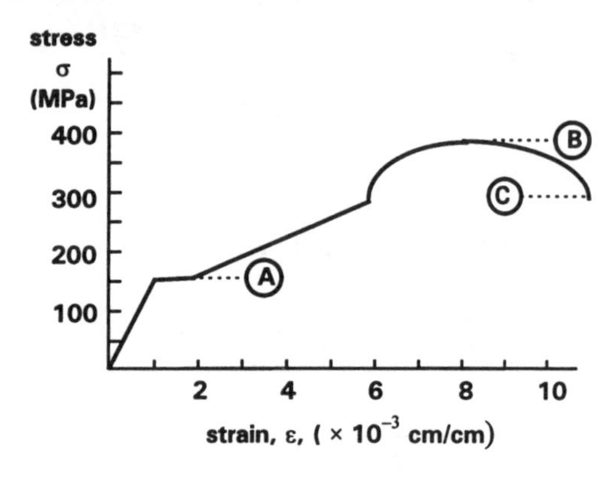

NUCLEAR POWER SYSTEMS–22

Point A represents which of the following?

(A) nil-ductility transition (NDT)
(B) ultimate tensile strength (UTS)
(C) work hardening point
(D) yield point (YP)

Solution:

The nil-ductility transition (NDT) is a temperature below which brittle fracture (also called cleavage fracture) tends to occur and above which ductile fracture (also called dimple fracture) tends to occur. The ultimate tensile stress (UTS) is the maximum point in the stress-strain curve. In this case, point B. It represents the maximum stress a metal can sustain without failure. Work hardening occurs beyond point A due to internal structural changes.

The point itself is referred to as the yield point (YP) and represents the point of transition between elastic and plastic deformation.

Answer is D.

NUCLEAR POWER SYSTEMS–23

What is the value of Young's modulus?

(A) 1.50×10^5 Pa
(B) 7.50×10^{10} Pa
(C) 1.50×10^{11} Pa
(D) 4.00×10^{11} Pa

Solution:

Young's modulus, E, is defined in Hooke's law as

$$E = \frac{\sigma}{\varepsilon}$$

σ = stress (Pa)
ε = strain[38]

In other word's, E is the slope of the stress-strain curve in the elastic region (the straight line portion from the origin to point A).

Substituting values obtained from the curve gives

$$\begin{aligned} E &= \frac{\sigma}{\varepsilon} \\ &= \frac{150 \times 10^6 \text{ Pa}}{0.001} \\ &= \boxed{1.50 \times 10^{11} \text{ Pa}} \end{aligned}$$

Answer is C.[39]

---

[37] This is an engineering or nominal stress-strain curve and is based on the original dimensions of the metal. True stress and strain depend on the actual dimensions at a given instant. See the materials portions of the various texts in the Reference section for additional information.

[38] Strain is technically dimensionless, but it represents $\Delta L/L_o$ (i.e., the change in length referenced to the original length).

[39] A typical value for Young's modulus for many steels is 2×10^{11} Pa.

NUCLEAR POWER SYSTEMS–24

A certain reactor support assembly is constructed of steel and constrained along its x-axis as shown.

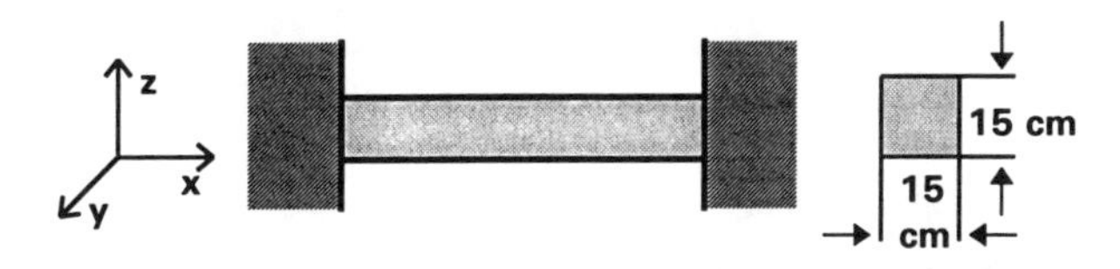

The assembly changes temperature from 20°C to 315°C during a startup from cold conditions. From a materials handbook, the coefficient of linear thermal expansion is $17 \times 10^{-6}\ \text{K}^{-1}$.

What is the compressive thermal stress on the assembly during hot operation?

(A) 3.8×10^{8} Pa
(B) 7.5×10^{8} Pa
(C) 2.0×10^{9} Pa
(D) 1.7×10^{11} Pa

Solution:

Hooke's law in equation form is

$$E = \frac{\sigma}{\varepsilon}$$

σ = stress (Pa)
ε = strain

Rearranging to solve for the stress gives

$$\sigma = E\varepsilon$$

As the material is unconstrained in the y and z directions, solve for the x direction values only, thus

$$\sigma_x = E\varepsilon_x$$

The strain in the x direction can be determined from

$$\varepsilon_x = \alpha_x \Delta T$$

α = coefficient of linear thermal expansion[40]
ΔT = temperature difference

Substituting gives

$$\begin{aligned}\varepsilon_x &= \alpha_x \Delta T\\ &= \left(17 \times 10^{-6}\ \text{K}^{-1}\right)(315^\circ\text{C} - 20^\circ\text{C})\end{aligned}$$

Noting that a temperature difference of one degree Celsius equals a temperature difference of one degree Kelvin gives[41]

$$\begin{aligned}\varepsilon_x &= \alpha_x \Delta T\\ &= \left(17 \times 10^{-6}\ \text{K}^{-1}\right)(315^\circ\text{C} - 20^\circ\text{C})\\ &= \left(17 \times 10^{-6}\ \text{K}^{-1}\right)(295^\circ\text{C}) = \left(17 \times 10^{-6}\ \text{K}^{-1}\right)(295\text{K})\\ &= 5.02 \times 10^{-3}\end{aligned}$$

Substituting this value and the previously determined value of Young's modulus from Prob. 23 into the rearranged Hooke's law equation gives

$$\begin{aligned}\sigma &= E\varepsilon\\ &= \left(1.5 \times 10^{11}\ \text{Pa}\right)\left(5.02 \times 10^{-3}\right)\\ &= \boxed{7.5 \times 10^{8}\ \text{Pa}}\end{aligned}$$

Answer is B.[42]

[40] Assumed here to be identical for all three directions (i.e., the material is isotropic).
[41] It is important to note that the temperatures are not equal; it is the temperature differences, the ΔTs, that are equal.
[42] In this problem the thermal stress was the result of a plant heatup. Such a stress can also occur due to radiation induced heating and would be termed swelling. See the Nuclear Radiation section for additional information.

NUCLEAR POWER SYSTEMS–25

A simplified diagram of a natural circulation boiling-water reactor (BWR) is shown.

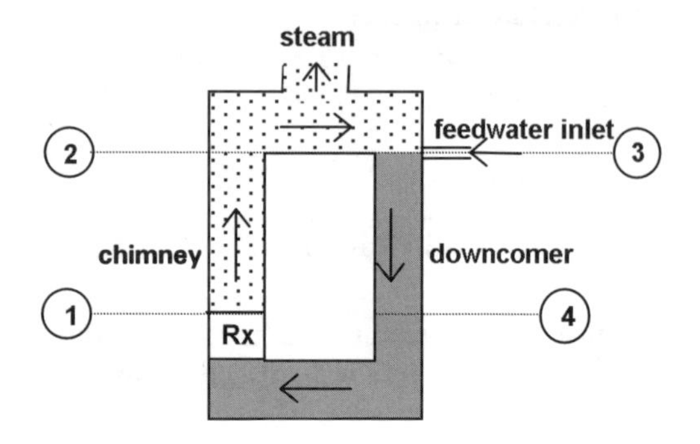

[steam water mixture]

[water]

Plant Data
chimney height = 10 m
steam pressure = 7.0 MPa
steam exit quality = 13%

The following three assumptions are used for this approximate calculation: (1) friction losses outside the core are negligible, (2) the system is isothermal, and (3) no heat transfer takes place to or from the surrounding environment.[43]

What is the approximate driving pressure available for natural circulation?[44]

(A) 70 kPa
(B) 75 kPa
(C) 80 kPa
(D) 85 kPa

[43] Clearly the isothermal condition does not occur. Additional data are required for a more accurate analysis. The incoming feed water will not be at the temperature of a saturated liquid regardless of how complete the mixing in the downcomer region.
[44] This is similar to the type of calculation necessary to determine the circulation head in a steam generator of a PWR plant.

Solution:

A review of the First Law of Thermodynamics for a closed system results in an energy balance for this system of

$$\Delta \text{PE} + \Delta \text{KE} + \Delta U = Q - W$$

PE = potential energy
KE = kinetic energy
U = internal energy
Q = heat added *to* the system
W = work done *by* the system

Using assumption (3) which is that no heat transfer occurs to or from the surrounding environment lets $Q = 0$. No pump exists in this natural circulation system, so $W = 0$. Writing the equation in specific energy terms (i.e., per unit mass) gives

$$\Delta \text{pe} + \Delta \text{ke} + \Delta u = 0$$

Expand the individual terms.

$$(z_f - z_0)g + \left(\frac{1}{2}\right)(\text{v}_f^2 - \text{v}_0^2) + (p_f - p_0)\upsilon = 0$$

The subscript "*f*" refers to the final condition; the subscript "0" (zero) refers to the initial condition. The velocity is represented by "v" and the specific volume by "υ."[45]

Since the velocity differences are zero in this closed system under steady state conditions, the equation is further simplified to

$$(z_f - z_0)g + (p_f - p_0)\upsilon = 0$$

Writing the equation from points 1 to 2 gives

$$\text{(i)} \quad (z_2 - z_1)g + (p_2 - p_1)\upsilon_{21} = 0$$

Writing the equation from points 3 to 4 gives

$$(z_4 - z_3)g + (p_4 - p_3)\upsilon_{43} = 0$$

[45] This equation is the extended Bernoulli equation or the mechanical energy equation.

By observation, points 2 and 3 are identical. Therefore, change the subscript from 3 to 2.

$$\text{(ii)}\quad (z_4 - z_2)g + (p_4 - p_2)\upsilon_{42} = 0$$

The pressure drop from points 4 to 1 represents the driving pressure for natural circulation. The equations can now be solved assigning the height at $z_1 = z_4 = 0$ and using the information that $p_2 = p_3 = 7.0$ MPa.

To ease this process, note that $\upsilon = 1/\rho$. Therefore Eq. (i) can be rewritten as

$$\text{(iii)}\quad (z_2 - z_1)g\rho_{21} + (p_2 - p_1) = 0$$

Equation (ii) can be rewritten as

$$\text{(iv)}\quad (z_4 - z_2)g\rho_{42} + (p_4 - p_2) = 0$$

Add Eqs. (iii) and (iv) to obtain

$$(z_2 - z_1)g\rho_{21} + (z_4 - z_2)g\rho_{42} + (p_4 - p_1) = 0$$

Rearranging for the desired pressure drop gives

$$\begin{aligned}\text{(v)}\quad (p_4 - p_1) &= -(z_2 - z_1)g\rho_{21} - (z_4 - z_2)g\rho_{42}\\ &= (z_1 - z_2)g\rho_{21} + (z_2 - z_4)g\rho_{42}\end{aligned}$$

At this point, other than the desired quantity $(p_4 - p_1)$, the items needing to be calculated are ρ_{21} and ρ_{42}.

For ρ_{21}, a two-phase mixture at saturation conditions with a quality of 13% at 7.0 MPa, results in a specific volume of

$$\begin{aligned}\upsilon_{21} &= 0.87\upsilon_g + 0.13\upsilon_f\\ &= (0.87)\left(27.37\ \frac{\text{cm}^3}{\text{g}}\right) + (0.13)\left(1.3513\ \frac{\text{cm}^3}{\text{g}}\right)\\ &= 23.99\ \text{cm}^3/\text{g}\end{aligned}$$

The information for υ_g and υ_f were obtained from Table 1.5 in App. 1.B. Use this calculated value of the specific volume to obtain the following density.

$$\rho_{21} = \frac{1}{\upsilon_{21}} = \frac{1}{23.99\ \frac{\text{cm}^3}{\text{g}}} = 0.04\ \text{g/cm}^3$$

Since $\upsilon_{42} = \upsilon_f$ at 7.0 MPa,

$$\upsilon_{42} = 1.3513\ \text{cm}^3/\text{g}$$

Thus,

$$\rho_{42} = \frac{1}{\upsilon_{42}} = \frac{1}{1.3513\ \frac{\text{cm}^3}{\text{g}}} = 0.74\ \text{g/cm}^3$$

Recall $z_1 = z_4 = 0$ by definition, and $z_2 = 10$ m because this is the chimney height. Substituting in Eq. (v) gives

$$\begin{aligned}(p_4 - p_1) &= (z_1 - z_2)g\rho_{21} + (z_2 - z_4)g\rho_{42}\\ &= (0\ \text{m} - 10\ \text{m})\left(9.8\ \frac{\text{m}}{\text{s}^2}\right)\left(0.04\ \frac{\text{g}}{\text{cm}^3}\right)\\ &\quad + (10\ \text{m} - 0\ \text{m})\left(9.8\ \frac{\text{m}}{\text{s}^2}\right)\left(0.74\ \frac{\text{g}}{\text{cm}^3}\right)\\ &= 68.60\ \frac{\text{m}^2\cdot\text{g}}{\text{s}^2\cdot\text{cm}^3}\end{aligned}$$

The units were expected to be in terms of pressure (Pa), which they are, but unrecognizable at this point. One can simply change to the expected unit; but, a more thorough approach is to convert the above units. Thus,

$$\begin{aligned}68.60\ \frac{\text{m}^2\cdot\text{g}}{\text{s}^2\cdot\text{cm}^3} &= \left(68.60\ \frac{\text{m}^2\cdot\text{g}}{\text{s}^2\cdot\text{cm}^3}\right)\left(\frac{1\ \text{kg}}{10^3\ \text{g}}\right)\left(\frac{10^2\ \text{cm}}{1\ \text{m}}\right)^3\\ &= \left(68.60\times10^3\ \frac{\text{kg}}{\text{s}^2\cdot\text{m}}\right)\left(\frac{1\ \text{N}}{1\ \frac{\text{kg}\cdot\text{m}}{\text{s}^2}}\right)\\ &= \left(68.60\times10^3\ \frac{\text{N}}{\text{m}^2}\right)\left(\frac{1\ \text{Pa}}{1\ \frac{\text{N}}{\text{m}^2}}\right)\\ &= \boxed{68.60\times10^3\ \text{Pa}\quad(70\ \text{kPa})}\end{aligned}$$

Answer is A.

NUCLEAR POWER SYSTEMS–26

Reactor control (i.e., the method of changing the neutron flux) involves the use of control rods. Additionally, both fixed burnable poisons and those in solution in the primary coolant are used to control the flux. A table of three common materials used in such applications and their properties follows.

designation	material
I	boron (B)
II	hafnium (Hf)
III	cadmium (Cd)

Properties

A—fairly large neutron absorption cross section over multiple isotopes

B—neutron absorption cross section in the thousands of barns

C—low melting point

D—resistant to corrosion in high temperature water

E—utilized in primary systems' coolant as a chemical shim

F—capable of decreasing reactor power level rapidly in its normal form as a control rod

Match the material with its properties. (The material may have more than one property associated with it and the properties may be used more than once.)

(A) I AC II BAC III BCD
(B) I AF II BCD III BCE
(C) I BF II AEF III BDE
(D) I BE II ADF III BCF

Solution:

Boron has a neutron absorption cross section of approximately 3800 barns (B) and is often used as a *chemical shim* (E) (i.e., it is dissolved in the primary coolant and the concentration varied in order to vary the *reactivity*).

Hafnium has a fairly large neutron absorption cross section, approximately 100 barns, over multiple isotopes (A). It is resistant to corrosion in high temperature water (D). Both of these properties combine to make hafnium desirable as a control rod (F).

Cadmium has an extremely high neutron absorption cross section of approximately 20 000 barns (B) but has the disadvantage of having a relatively low melting point of 321°C (610°F) (C). The melting point problem is overcome by alloying the material with silver and indium and using this combination as a control rod (F).

Therefore, the appropriate combinations are

I-BE II-ADF III-BCF

Answer is D.

8 NUCLEAR FUEL MANAGEMENT: SOLUTIONS

Nuclear Fuel Management (NFM) problems 1–5 are based on the following information and equations.

The uranium required as fuel for PWRs is mined as U_3O_8, purified using solvent extraction processes to remove impurities,[1] and then converted using chemical processes to uranium hexafluoride (UF_6).[2] The UF_6 is then enriched, usually to a nominal value of 3 w/o [weight percent] of U-235, using a gaseous diffusion process.[3]

The gaseous diffusion process is explained by two major equations.

$$F = P + W$$

$$x_f F = x_p P + x_w W$$

x_f = weight fraction of U-235 in the feed material[4]
x_p = weight fraction of U-235 in the product (this is the desired enrichment)
x_w = weight fraction of U-235 in the waste stream (this is the depleted uranium); also called the tails assay
F = number of kilograms of feed material, per unit time
P = number of kilograms of product enriched, per unit time
W = number of kilograms of uranium in the waste stream, per unit time[5]

[1] In the U.S., the PUREX process or a method using uranium peroxide ($UO_4 \cdot 2H_2O$) is utilized.
[2] The process used is dry hydroflour or wet solvent extraction.
[3] Other methods possible include the centrifuge, separation nozzle and ALVIS (Atomic Vapor Laser Isotope Separation).
[4] Note that this is a fraction, not a percent. Thus, natural uranium, at present, consists of 0.711% U-235. Therefore, x_f is 0.00711.
[5] F, P, and W are defined as mass flow rate (mass/time). These terms are frequently referred to as the "total" mass, without reference to time. No error is introduced during calculations if one is consistent.

One quantity derived from the gaseous diffusion process equations is the *feed factor*.

$$\frac{F}{P} = \frac{x_p - x_w}{x_f - x_w}$$

Another quantity derived from these equations is the *waste factor*.

$$\frac{W}{P} = \frac{x_p - x_f}{x_f - x_w} = \frac{F}{P} - 1$$

NUCLEAR FUEL MANAGEMENT–1

The feed factor F/P is a tabulated quantity which indicates the number of kilograms of uranium needed as feed for the enrichment process per kilogram of product. If a utility specifies an enrichment x_p of 3% with a tails assay of 0.3%, what is the feed factor?

(A) 0.04
(B) 0.40
(C) 3.30
(D) 6.60

Solution:

The applicable formulas are

$$\text{(i)} \quad F = P + W$$

$$\text{(ii)} \quad x_f F = x_p P + x_w W$$

Although the formula for the feed factor was given, its instructive to obtain it from Eqs. (i) and (ii). Rearranging Eq. (i) gives

$$W = F - P$$

Substituting this expression for W into Eq. (ii) gives

$$x_f F = x_p P + x_w (F - P)$$

Expand the term in parenthesis and rearrange to solve for the feed factor.

$$x_f F = x_p P + x_w F - x_w P$$
$$F(x_f - x_w) = P(x_p - x_w)$$
$$\frac{F}{P} = \frac{x_p - x_w}{x_f - x_w}$$

A second method involves solving equation (ii) for W giving

$$x_f F = x_p P + x_w W$$
$$x_w W = x_f F - x_p P$$
$$W = \frac{x_f F - x_p P}{x_w}$$

Substitute this expression for W into Eq. (i).

$$F = P + W$$
$$F = P + \left(\frac{x_f F - x_p P}{x_w}\right)$$
$$x_w F = x_w P + x_f F - x_p P$$
$$x_w F - x_f F = x_w P - x_p P$$
$$F(x_w - x_f) = P(x_w - x_p)$$
$$\frac{F}{P} = \frac{x_w - x_p}{x_w - x_f}$$

Note: Although this second result is certainly valid mathematically, it results in a negative numerator and denominator. Therefore, the first form of the feed factor is normally used. Thus,

$$\frac{F}{P} = \frac{x_p - x_w}{x_f - x_w}$$

The enrichment is given as 3%, thus,

$$x_p = 0.03$$

Natural uranium consists of 0.711% U-235 by weight, thus,

$$x_f = 0.00711$$

The tails assay (i.e., the waste stream U-235 weight fraction) is specified as 0.3%. Thus,

$$x_f = 0.003$$

Substituting these values into the feed factor formula gives[6]

$$\frac{F}{P} = \frac{x_p - x_w}{x_f - x_w}$$
$$= \frac{0.03 - 0.003}{0.00711 - 0.003}$$
$$= \boxed{6.57 \quad (6.60 \text{ kg feed/kg product})}$$

Answer is D.[7]

NUCLEAR FUEL MANAGEMENT–2

During annual refueling a batch of fuel assemblies consisting of one-third of the 33 000 kg of 2.7 w/o (i.e., weight percent) U-235 will be replaced.

The cost of natural uranium is $40/kg. The cost of *conversion* from U_3O_8 to UF_6 is $5/kg and suffers a material loss of 0.5% uranium. Ignore any fabrication losses. The tails assay for the replacement fuel is 0.3%.

What is the cost through conversion of the natural uranium that must be provided as feed to an enrichment plant to provide for the year's refueling needs?

(A) $65,000
(B) $264,000
(C) $2,900,000
(D) $8,700,000

[6] In this case, percentage values could be used in the feed factor formula and the answer shown would have been identical. It is recommended one use fractions at all times in these types of calculations to avoid errors induced by using percentages when calculating the separative work unit (SWU) factor S as shown in other problems.
[7] Technically the answer is unitless. Carrying the indicated units helps clarify extended calculations.

Solution:

First, the price of uranium through conversion to UF_6 and ignoring fabrication losses is

$$P_{n\to c} = \left(\frac{P_n}{1-l_c} + P_c\right)\left(\frac{F}{P}\right)$$

$P_{n\to c}$ = price of natural uranium including conversion
P_n = price of natural uranium
P_c = price of conversion
F/P = feed factor in units of kg-feed required/kg-product desired

Substituting the given into the feed factor formula gives

$$\begin{aligned}\frac{F}{P} &= \frac{x_p - x_w}{x_f - x_w}\\ &= \frac{0.027 - 0.003}{0.00711 - 0.003}\\ &= 5.84 \quad (5.84 \text{ kg-feed/kg-product})\end{aligned}$$

Substituting into the price formula gives

$$\begin{aligned}P_{n\to c} &= \left(\frac{P_n}{1-l_c} + P_c\right)\left(\frac{F}{P}\right)\\ &= \left(\frac{40\frac{\$}{\text{kg}}}{1-0.005} + 5\frac{\$}{\text{kg}}\right)(5.84)\\ &= 263.97 \text{ \$/kg}\end{aligned}$$

The desired result can now be obtained from the following.

$$\begin{aligned}\text{Total Cost} &= \left(263.97\frac{\$}{\text{kg}}\right)(11000 \text{ kg})\\ &= \boxed{\$2{,}903{,}670 \quad (\$2{,}900{,}000)}\end{aligned}$$

Answer is C.[8]

[8] Obviously more extensive analysis would be required to account for inflation, the cost of any borrowed funds (i.e., interest) the effect of taxes and so on. Use any good engineering economics text to review the principles of these and other effects. See the Reference section for an example of such a text.

NUCLEAR FUEL MANAGEMENT–3

A separative work unit (SWU) is a quantity directly related to the resources required to perform the enrichment to the desired level of x_p, given values of x_f and x_w, and is given by

$$\text{SWU} = \left(PV(x_p) + WV(x_w) - FV(x_f)\right)T$$

SWU = separative work unit in kg or kg-SWU[9]
$V(x_i)$ = separation potentials with

$$V(x_i) = (2x_i - 1)\ln\left(\frac{x_i}{1-x_i}\right)$$

T = time period, usually a year

Assume the nominal one year time frame. What is the number of kilograms of natural uranium that must be provided as feed to the enrichment plant if one requests 100 000 kg[10] at 2.8% in U-235 enriched uranium with a known tails assay of 0.25%? Also, what is the number of SWUs needed for the separation?

(A) 100×10^3 kg; 3×10^6 kg-SWU
(B) 103×10^3 kg; 3×10^6 kg-SWU
(C) 48×10^3 kg; 340×10^3 kg-SWU
(D) 553×10^3 kg; 340×10^3 kg-SWU

Solution:

The total feed uranium required can be calculated from the feed factor.

$$\frac{F}{P} = \frac{x_p - x_w}{x_f - x_w}$$

$$F = P\left(\frac{x_p - x_w}{x_f - x_w}\right)$$

[9] The SWU portion of the unit shown is not a unit per se. It's used for convenience in calculating the separative work unit factor S as will be shown in a later problem.
[10] This refers to the kg of enriched product (UF_6), not just uranium.

Substituting the given information gives the value for the total feed required.

$$F = P\left(\frac{x_p - x_w}{x_f - x_w}\right)$$
$$= (100\,000 \text{ kg product})\left(\frac{0.028 - 0.0025}{0.00711 - 0.0025}\right)$$
$$= \boxed{553 \times 10^3 \text{ kg feed uranium}}$$

Note: Total values were used instead of per unit time values.

Consider the SWU formula.

$$\text{SWU} = \left(PV(x_p) + WV(x_w) - FV(x_f)\right)T$$

P is given as 100 000 kg; F is calculated; T is given as one year but should be assumed to be one year unless stated otherwise, and the product P specified is assumed to be produced over that same year time frame.

To find W use

$$F = P + W$$
$$W = F - P$$
$$= 553.15 \times 10^3 \text{ kg feed} - 100.00 \times 10^3 \text{ kg product}$$
$$= 453.15 \times 10^3 \text{ kg}$$

Recall that the "feed" and "product" terms on the units are for convenience.

The separation potential for the product is

$$V(x_p) = (2x_p - 1)\ln\left(\frac{x_p}{1 - x_p}\right)$$
$$= \left((2)(0.028) - 1\right)\ln\left(\frac{0.028}{1 - 0.028}\right) = 3.35$$

For the waste,

$$V(x_w) = (2x_w - 1)\ln\left(\frac{x_w}{1 - x_w}\right)$$
$$= \left((2)(0.0025) - 1\right)\ln\left(\frac{0.0025}{1 - 0.0025}\right) = 5.96$$

For the feed,

$$V(x_f) = (2x_f - 1)\ln\left(\frac{x_f}{1 - x_f}\right)$$
$$= \left((2)(0.00711) - 1\right)\ln\left(\frac{0.00711}{1 - 0.00711}\right) = 4.87$$

Substituting gives[11]

$$\text{SWU} = \left(PV(x_p) + WV(x_w) - FV(x_f)\right)$$
$$= (100\,000 \text{ kg})(3.35) + (453.15 \times 10^3 \text{ kg})(5.96)$$
$$- (553.15 \times 10^3 \text{ kg})(4.87)$$
$$= \boxed{341.93 \times 10^3 \text{ kg-SWU} \quad (340 \times 10^3 \text{ kg-SWU})}$$

Answer is D.

NUCLEAR FUEL MANAGEMENT–4

While the SWU provides information on the resources required to enrich the desired amount of product, it varies with the amount of product and the time frame. To standardize, a SWU factor S is defined as the number of SWUs per unit of product.[12] In equation form,

$$S = \frac{\text{SWU}}{PT} = V(x_p) + \left(\frac{W}{P}\right)V(x_w) - \left(\frac{F}{P}\right)V(x_f)$$

P is the number of kilograms per unit time and T is the time, usually one year.[13]

What is the SWU factor for a 3% enrichment of natural uranium using a 0.25% tails assay?

(A) 2.715
(B) 3.780
(C) 3.811
(D) 30.588

[11] The time period T is one year and is not shown in the SWU formula in order to use P, W, and F in terms of their total values (i.e., kg). If T had been used, P, W, and F would need to be specified as kg/yr.
[12] The factor S is technically dimensionless. Its units are kg-SWU/kg. The SWU factor S provides the number of SWUs required per kilogram of product, not per gram or lbm.
[13] The SWU factor S is usually tabulated along with values of the feed factor F/P.

Solution:

The formula for the SWU factor (S) is

$$S = \frac{\text{SWU}}{PT} = V(x_p) + \left(\frac{W}{P}\right)V(x_w) - \left(\frac{F}{P}\right)V(x_f)$$

Solving for the separation potentials gives

$$\begin{aligned} V(x_p) &= (2x_p - 1)\ln\left(\frac{x_p}{1-x_p}\right) \\ &= ((2)(0.03) - 1)\ln\left(\frac{0.03}{1-0.03}\right) \\ &= 3.268 \end{aligned}$$

$$\begin{aligned} V(x_w) &= (2x_w - 1)\ln\left(\frac{x_w}{1-x_w}\right) \\ &= ((2)(0.0025) - 1)\ln\left(\frac{0.0025}{1-0.0025}\right) \\ &= 5.959 \end{aligned}$$

$$\begin{aligned} V(x_f) &= (2x_f - 1)\ln\left(\frac{x_f}{1-x_f}\right) \\ &= ((2)(0.00711) - 1)\ln\left(\frac{0.00711}{1-0.00711}\right) \\ &= 4.869 \end{aligned}$$

The feed factor is

$$\begin{aligned} \frac{F}{P} &= \frac{x_p - x_w}{x_f - x_w} \\ &= \frac{0.03 - 0.0025}{0.00711 - 0.0025} \\ &= 5.965 \end{aligned}$$

Therefore, the waste factor is

$$\begin{aligned} \frac{W}{P} &= \frac{F}{P} - 1 \\ &= 5.965 - 1 \\ &= 4.965 \end{aligned}$$

Substituting the calculated and given values into the SWU factor equation gives

$$\begin{aligned} S &= V(x_p) + \left(\frac{W}{P}\right)V(x_w) - \left(\frac{F}{P}\right)V(x_f) \\ &= 3.268 + (4.965)(5.959) - (5.965)(4.869) \\ &= \boxed{3.811 \quad (3.811\ \text{kg-SWU/kg})} \end{aligned}$$

Answer is C.

NUCLEAR FUEL MANAGEMENT–5

The following information is taken from the technical documentation provided by an enrichment facility:

P_u = price of natural uranium = $50/kg
P_c = price of conversion = $8/kg
P_{SWU} = price of separative work units = $90/SWU
l_c = conversion loss = 0.4%
x_w = tails assay = 0.3%

A plant requires 2.7 w/o (i.e., weight percent) enriched uranium for an upcoming refueling. What is the price per kilogram of enriched uranium required by the plant? Ignore fabrication losses.

(A) $300/kg
(B) $600/kg
(C) $800/kg
(D) $1100/kg

Solution:

Note in this problem the cost of the uranium is from mining through enrichment. The applicable formula is[14]

$$P_E = \left(\frac{P_u}{1-l_c} + P_c\right)\left(\frac{F}{P}\right) + P_{SWU}S$$

P_E = price of enriched uranium per kilogram
S = SWU factor

Two components of the equation for P_E are not provided, the feed factor (F/P) and the SWU factor (S).

[14] Memorization of the formula is not required. One can derive it from an analysis of the units keeping in mind at what point in the mining-conversion-enrichment-fabrication process one desires the cost.

Calculating the feed factor gives

$$\frac{F}{P} = \frac{x_p - x_w}{x_f - x_w} = \frac{0.027 - 0.003}{0.00711 - 0.003} = 5.839 \quad (5.839 \text{ kg feed/kg product})$$

The SWU factor formula is

$$S = \frac{\text{SWU}}{PT} = V(x_p) + \left(\frac{W}{P}\right)V(x_w) - \left(\frac{F}{P}\right)V(x_f)$$

Calculating the separation potentials first gives[15]

$$V(x_p) = (2x_p - 1)\ln\left(\frac{x_p}{1 - x_p}\right) = ((2)(0.027) - 1)\ln\left(\frac{0.027}{1 - 0.027}\right) = 3.391$$

$$V(x_w) = (2x_w - 1)\ln\left(\frac{x_w}{1 - x_w}\right) = ((2)(0.003) - 1)\ln\left(\frac{0.003}{1 - 0.003}\right) = 5.771$$

$$V(x_f) = (2x_f - 1)\ln\left(\frac{x_f}{1 - x_f}\right) = ((2)(0.00711) - 1)\ln\left(\frac{0.00711}{1 - 0.00711}\right) = 4.869$$

The waste factor is

$$\frac{W}{P} = \frac{x_p - x_f}{x_f - x_w} = \frac{F}{P} - 1 = 5.839 - 1 = 4.839$$

Substituting into the equation for the SWU factor gives

$$S = \frac{\text{SWU}}{PT} = V(x_p) + \left(\frac{W}{P}\right)V(x_w) - \left(\frac{F}{P}\right)V(x_f) = 3.391 + (4.839)(5.771) - (5.839)(4.869) = 2.888$$

Substituting into the formula for the price of enriched uranium gives[16]

$$P_E = \left(\frac{P_u}{1 - l_c} + P_c\right)\left(\frac{F}{P}\right) + P_{SWU}S$$

$$= \left(\frac{50\,\frac{\$}{\text{kg}}}{1 - 0.004} + 8\,\frac{\$}{\text{kg}}\right)\left(5.839\,\frac{\text{kg feed}}{\text{kg product}}\right) + \left(90\,\frac{\$}{\text{SWU}}\right)\left(2.888\,\frac{\text{kg-SWU}}{\text{kg}}\right)$$

$$= \boxed{\$599.75/\text{kg product} \quad (\$600/\text{kg})}$$

Answer is B .

[15] Note that the separation potential for the feed, $V(x_f)$, remains unchanged at 4.869 in different problems as it only depends on the weight percent (w/o) of U-235 in natural uranium (0.711%).

[16] Be very careful with units here. An understanding of the process makes the equation easier to understand. Note that kg feed is identical to the kg unit in the first term and that kg-SWU and SWU are identical units in the second term. Further, the kg unit in the second term represents the product and is equivalent to the unit kg product.

Nuclear Fuel Management (NFM) problems 6 and 7 are based on the following information and illustration.

A common fabrication process for pellet type fuel is one which starts with enriched uranium hexafluoride (UF_6) supplied in a high pressure cylinder as a solid. It is then heated, causing the UF_6 to sublime. The gas bubbles through water, forms UO_2F_2, and is mixed with ammonia water to precipitate the uranium as ammonium diuranate $(NH_4)_2U_2O_2$. The precipitate is dried at high temperature (*calcined*) to form U_3O_8. Hydrogen is used to further reduce the compound to uranium dioxide, UO_2. This is ground into a fine powder added to an adhesive agent and then pressed into a cylindrical pellet. The pellets are then sintered near the melting point to cause densification to occur. The pellets are then loaded into a fuel rod assembly such as shown.

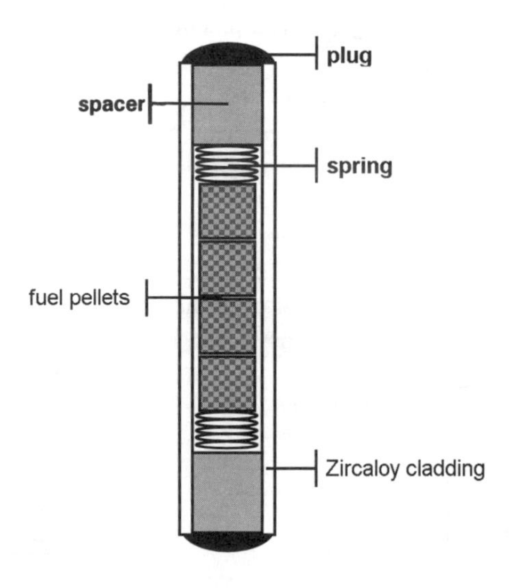

NUCLEAR FUEL MANAGEMENT–6

Which of the following is not a reason for the use of uranium dioxide (UO_2) as a form of nuclear fuel?

(A) good thermal conductivity
(B) chemically stable
(C) structurally stable
(D) oxygen's low neutron capture cross section

Solution:

Consider each item in turn.

First, the thermal conductivity. Using information from Table 2.1 in App. 2.A, at 650K (an approximate fuel centerline temperature) the thermal conductivity of UO_2 is approximately 4.6 W/m·K. By contrast, Zircaloy is 17.0 W/m·K, stainless steel is 19.0 W/m·K and carbon steel is 41.0 W/m·K. Thus, UO_2 thermal conductivity is low compared with other commonly used reactor plant materials and is not a positive attribute for its use as a form of nuclear fuel.

Second, the chemical stability. Uranium (U) metal reacts quickly with air to oxidize. Additionally, it reacts with water to evolve hydrogen, often quite severely. Uranium oxide (UO_2) on the other hand is relatively stable in terms of chemical reactivity and indeed suffers no reaction with high temperature water.

Third, the structural stability. Although uranium oxide's strength is inferior to that of uranium metal and is not a simple material, it is more structurally stable than uranium metal which undergoes the following changes or phases.

phase	structure	temperature
α	orthorhombic	ambient–600°C
β	tetragonal	660°C–700°C
γ	body centered cubic	760°C–Melting Point

Such phase changes present significant design challenges for the reactor and mechanical engineers using this material.

Fourth, the thermal neutron absorption cross section (σ_z). From data in Table 2.2 of App. 2.A, the value for σ_z for oxygen is 0.00027 barns; for zirconium it's 0.185 barns; and for iron it's 2.55 barns.[17] Thus, the neutron absorption cross section for oxygen is low which would make it useful as part of the nuclear fuel.

In conclusion, the only trait discussed that is not a reason for the use of UO_2 as a nuclear fuel is that the thermal conductivity is low.

Answer is A.

[17] A barn is 10^{-24} cm^2.

NUCLEAR FUEL MANAGEMENT–7

Individualized fuel pellets are "cupped" as shown.

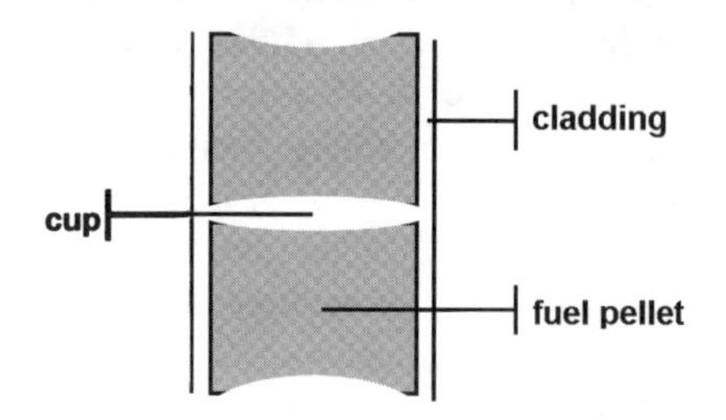

The purpose of the cup is to

(A) allow space for fission products
(B) increase mechanical compression on the outer portion of the pellet to prevent movement
(C) allow for ease of handling during subsequent refueling
(D) trap hydrogen away from Zircaloy cladding to prevent formation of ZrH_2

Solution:

Numerous nuclear fuel problems have been encountered, and for the most part, overcome.

Consider each of the items in reverse order.

First, the formation of hydrides (ZrH_2). The hydrogen alluded to in D is generated from the radiolysis of water in the following reaction.

$$H_2O \xrightarrow{\gamma} H + OH$$

Additionally, the hydrogen can be picked up by the pellet during pressing or grinding.

The hydrogen absorbed by the Zircaloy lattice can result in the formation of ZrH_2 which in large quantities causes embrittlement. Most of the hydrogen is generated outside the fuel assembly in reaction with a water moderator.[18] Further, the cups are not capable of sealing the hydrogen away from the cladding, even if most did originate inside the fuel assembly, due to the small size of the hydrogen.[19] Therefore, D is not the answer.

Second, pellets are not handled individually during refueling. Therefore C is not the answer.

Third, the required mechanical compression is supplied by the springs. Therefore, B is not the answer.

Consider other fuel problems.

Pellet/cladding interaction (PCI) is a form of stress corrosion cracking which is a failure of a metal under stress in a corrosive chemical environment. This is unrelated to the cups.

Fuel *densification* results when the UO_2 pellets' density increases with operation resulting in gaps in the fuel stack. Cladding can then collapse into the gaps creating the potential for leakage. The densification mentioned as part of the final fabrication process minimizes this effect. Further, pore specifications in the fuel pellets aid in minimizing densification problems. Again, none of this influences cup size.

Thermal expansion occurs due to the anisotropic properties of uranium. As the fuel pellet is irradiated, the heating results in an expansion as shown.

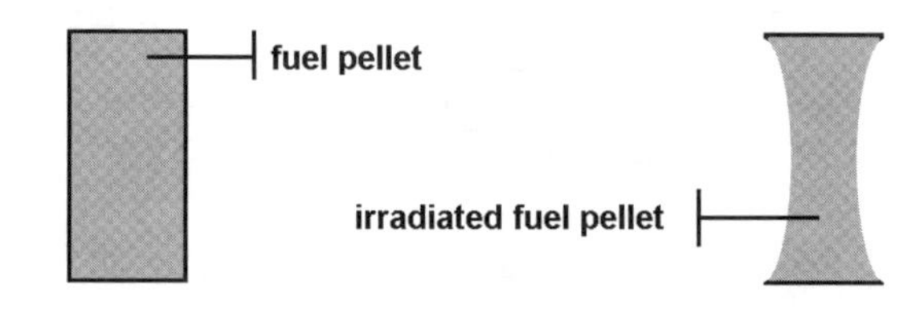

Cupping or dishing of the pellets largely eliminates this problem. However, thermal expansion is not one of the choices.

Fuel swelling results when gaseous fission products are formed filling the pores in the UO_2 pellets. The cups allow space for these gasses, mostly helium.

Answer is A.

[18] See the Nuclear Theory section for more information.

[19] Monatomic hydrogen due to its small size is capable of absorption into the Zircaloy metal matrix. Uranium dioxide pellets have the same difficulty in preventing its movement. See a materials text for further explanation of this important phenomenon. See the Reference section for an example of such a text.

NUCLEAR FUEL MANAGEMENT–8

During the fabrication of nuclear fuel a small loss, usually less than 1%, occurs. As a result of this loss, more uranium is required with a commensurate increase in the cost at each stage—mining, conversion, enrichment, and fabrication. Representative prices and losses follow.

P_u = price of natural uranium = \$60/kgU
P_c = price of conversion = \$10/kgU
P_{SWU} = price of separative work unit = \$100/SWU
P_f = price of fabrication[20] =\$250/kgU
l_c = conversion material loss = 0.2%
l_f = fabrication material loss = 0.5%

Enrichment will be of natural uranium to a 3.3 w/o (i.e., weight percent) U-235 with a tails assay of 0.3%.[21] What is most nearly the total cost per kg-uranium of the fabricated fuel (P_f)?

(A) \$800/kgU
(B) \$950/kgU
(C) \$1050/kgU
(D) \$1150/kgU

Solution:

Let the total cost per kgU be P_{ff}. Then

$$P_{ff} = \left(\left(\frac{P_u}{(1-l_c)(1-l_f)} + \frac{P_c}{1-l_f}\right)\left(\frac{F}{P}\right) + \left(\frac{P_{\text{SWU}}}{1-l_f}\right)S + P_f\right)$$

Calculate the feed factor.

$$\begin{aligned} \frac{F}{P} &= \frac{x_p - x_w}{x_f - x_w} \\ &= \frac{0.033 - 0.003}{0.00711 - 0.003} \\ &= 7.299 \quad (7.299 \text{ kg feed/kg product}) \end{aligned}$$

Recall that the waste factor is the feed factor minus one. Thus,

$$\begin{aligned} \frac{W}{P} &= \frac{x_p - x_f}{x_f - x_w} = \frac{F}{P} - 1 \\ &= 7.299 - 1 \\ &= 6.299 \end{aligned}$$

The SWU factor is

$$S = \frac{\text{SWU}}{PT} = V(x_p) + \left(\frac{W}{P}\right)V(x_w) - \left(\frac{F}{P}\right)V(x_f)$$

Calculating the separation potentials first gives

$$\begin{aligned} V(x_p) &= (2x_p - 1)\ln\left(\frac{x_p}{1-x_p}\right) \\ &= ((2)(0.033) - 1)\ln\left(\frac{0.033}{1-0.033}\right) \\ &= 3.155 \end{aligned}$$

$$\begin{aligned} V(x_w) &= (2x_w - 1)\ln\left(\frac{x_w}{1-x_w}\right) \\ &= ((2)(0.003) - 1)\ln\left(\frac{0.003}{1-0.003}\right) \\ &= 5.771 \end{aligned}$$

$$\begin{aligned} V(x_f) &= (2x_f - 1)\ln\left(\frac{x_f}{1-x_f}\right) \\ &= ((2)(0.00711) - 1)\ln\left(\frac{0.00711}{1-0.00711}\right) \\ &= 4.869 \end{aligned}$$

Substituting into the SWU factor equation gives

$$\begin{aligned} S &= \frac{\text{SWU}}{PT} = V(x_p) + \left(\frac{W}{P}\right)V(x_w) - \left(\frac{F}{P}\right)V(x_f) \\ &= 3.155 + (6.299)(5.771) - (7.299)(4.869) \\ &= 3.968 \end{aligned}$$

[20] The cost of fabrication includes the cost of transportation to the site.
[21] The assumption is that the feed uranium is natural uranium with a weight percent of 0.711. This does not have to be the case. Previously enriched uranium which has been used as fuel and then reprocessed will have a higher input percentage depending on the burnup, on the order of 1 w/o.

The total price of the fuel through fabrication and including fabrication losses is then

$$P_{ff} = \left(\frac{P_u}{(1-l_c)(1-l_f)} + \frac{P_c}{1-l_f} \right)\left(\frac{F}{P}\right) + \left(\frac{P_{\text{SWU}}}{(1-l_f)} \right) S + P_f$$

$$= \left(\frac{60\frac{\$}{\text{kgU}}}{(1-0.002)(1-0.005)} + \frac{10\frac{\$}{\text{kgU}}}{1-0.005} \right)\left(7.299\frac{\text{kg feed}}{\text{kg product}}\right)$$

$$+ \left(\frac{100\frac{\$}{\text{SWU}}}{1-0.005} \right)\left(3.968\frac{\text{kg-SWU}}{\text{kg}}\right) + 250\frac{\$}{\text{kgU}}$$

$$= \boxed{1163.17\ \$/\text{kgU} \quad (\$1150/\text{kgU})}$$

Answer is D.[22]

NUCLEAR FUEL MANAGEMENT–9

A 500 MW_e reactor was operated during a given year as shown.

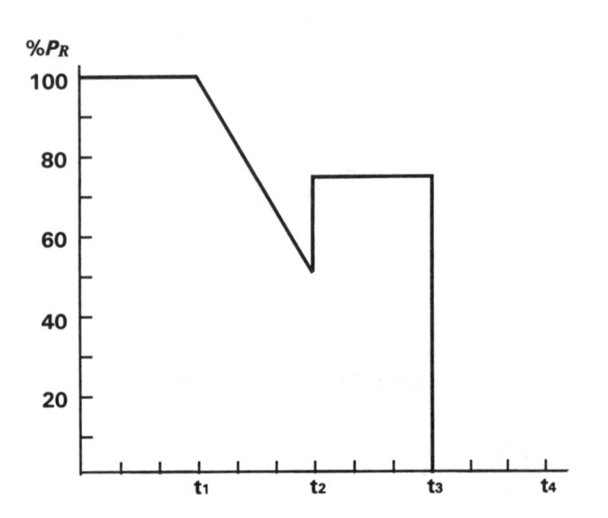

The rated power P_R is given as a percentage of full power. The year is broken into quarter periods t_1, t_2, t_3, and t_4.

What is the capacity factor in percent?

(A) 62.5%
(B) 72.5%
(C) 82.5%
(D) 92.5%

Solution:

The formula for the capacity factor (CF) is

$$\text{CF} = \int_0^T \frac{P(t)}{P_R T} dt$$

The *capacity factor* is defined as the percent of total electric power which could theoretically be produced during a specified period if the plant were operated at full power one hundred percent of the time.[23] The time period is T. For comparison of plants, the time period is normally taken to be one year.

Splitting the integral into quarter fragments and noting that the rated power P_R, the time period T, and the power during three of the four time periods are constants gives

$$\text{CF} = \int_0^T \frac{P(t)}{P_R T} dt$$

$$= \left(\frac{1}{P_R T}\right)\left(\int_{t_0}^{t_1} P(t)dt + \int_{t_1}^{t_2} P(t)dt + \int_{t_2}^{t_3} P(t)dt + \int_{t_3}^{t_4} P(t)dt \right)$$

$$= \left(\frac{1}{P_R T}\right)\left(P_{0,1}\int_{t_0}^{t_1} dt + \int_{t_1}^{t_2} P(t)dt + P_{2,3}\int_{t_2}^{t_3} dt + P_{3,4}\int_{t_3}^{t_4} dt \right)$$

$$= \left(\frac{1}{(500\ \text{MW})(365\ \text{days})} \right)\Big((500\ \text{MW})(91.25\ \text{days})$$
$$+ (0.75)(500\ \text{MW})(91.25\ \text{days}) + (0.75)(500\ \text{MW})(91.25\ \text{days})$$
$$+ (0\ \text{MW})(91.25\ \text{days})\Big)$$

$$= \boxed{0.625 \quad (62.5\%)}$$

Answer is A.[24]

[22] In the formula, kg-feed is equivalent to kgU in the first term. Also, kg-SWU is equivalent to SWU in the second term. The remaining terms, kgU, kg-product and kg all can be changed to kgU as is done for the final answer. There are other ways of handling these units (e.g., leaving the technically unitless feed factor F/P and the SWU factor S arbitrary units out of the equation and dealing only with kg and kgU). Use whatever makes sense to you based on an understanding of the terms.

[23] The maximum value is therefore one.

[24] Often the three month period is taken as 90 days.

The integrals were carried out by inspection. More formal methodology could be used but it would serve only to complicate the problem. Further, a closer inspection of the definition reveals that the sum of simple fractions representing the CF for each period can normally be calculated without regards to the fractions. For example,

$$\begin{aligned}\text{CF} &= \left(\frac{1}{4}\right)(1.0)+\left(\frac{1}{4}\right)(0.75)+\left(\frac{1}{4}\right)(0.75)+\left(\frac{1}{4}\right)(0.75)\\ &= 0.625 \quad (62.5\%)\end{aligned}$$

Either method is correct. Using the integral method would be necessary when the power history is more complex.

NUCLEAR FUEL MANAGEMENT–10

A 1000 MW_e plant with an efficiency of 30% is initially loaded with 90 000 kg of uranium dioxide (UO_2) fuel. The core operates at full power for one year and then shuts down for refueling. During refueling one-third of the fuel is replaced. What is the average burnup in units of megawatt days per metric ton (MWD/MTU) for the third of the core initially removed?[25]

(A) 4.6×10^3 MWD/MTU
(B) 13.8×10^3 MWD/MTU
(C) 15.3×10^3 MWD/MTU
(D) 45.6×10^3 MWD/MTU

Solution:

The burnup in equation form is

$$\text{BU} = \frac{TP_R CF}{\text{MTU}}$$

BU = burnup in MWD/MTU
T = time period in units of days
P_R = rated thermal power in MW
CF = capacity factor

Note that the power in the equation is the thermal power (i.e., MW_{th}). If using the electric power output (i.e., MW_e), remember to divide by the efficiency η_{th} in order to obtain the thermal power.

The burnup is based on the heavy metal (i.e., uranium and all its isotopes and anything of greater atomic number than 92).[26] The fuel is in the form of UO_2; the 90 000 kg therefore includes the oxygen mass. To calculate the metric tonnes of heavy metal use

$$\begin{aligned}\text{tonnes of heavy metal} &= m_{\text{fuel}}\left(\frac{m_{\text{U}}}{m_{\text{UO}_2}}\right)\\ &= (90\,000\ \text{kg fuel})\left(\frac{1\ \text{tonne}}{1000\ \text{kg}}\right)\\ &\quad\times\left(\frac{238\ \text{gU}}{\left(238\ \text{g}\,^{238}\text{U}+(2)(16\ \text{g}\,^{16}\text{O})\right)\text{gUO}_2}\right)\\ &= 79.3\ \text{MTU}\end{aligned}$$

The mass of U-238 is used for the mass of all the isotopes of uranium in the equation. The actual value is 238.03 amu (i.e., atomic mass units). The error introduced is less than one-tenth of 1%.[27]

The core operated for 365 days (i.e., one standard year) at 100% power giving a capacity factor of 1.0. Thus,

$$\begin{aligned}\text{BU} &= \frac{TP_R CF}{\text{MTU}}\\ &= \frac{T\left(\frac{P_e}{\eta_{\text{th}}}\right)(CF)}{\text{MTU}} = \frac{(365\ \text{days})\left(\frac{1000\ \text{MW}}{0.30}\right)(1.0)}{79.3\ \text{MTU}}\\ &= \boxed{15.3\times 10^3\ \text{MWD/MTU}}\end{aligned}$$

Answer is C.[28]

[25] This is a metric ton (i.e., 1000 kg of uranium) represented as "tonne" in some texts and MTU in others, as well as other variations. The reason for the unit change will be explained in the next footnote.

[26] The heavy metal in the core often includes plutonium that does account for a portion of the thermal energy generated. Since this is the initial core load, the fuel is assumed to be plutonium free and one calculates the burnup based on the total uranium present initially. Since plutonium is often included the MTU unit is changed to tonne, or just t.

[27] This assumption, that the atomic mass of natural uranium is 238 amu, is valid in many circumstances. Further, for those times when the uranium is enriched to increase the concentration of U-235, the assumption remains valid. This is for standard enrichments in power plants of approximately 3%. Obviously, the higher the enrichment, the less valid the assumption.

[28] The amount of uranium used in the calculation is the total amount in the core that experiences the burnup, not merely the one-third which was removed.

Nuclear Fuel Management (NFM) problems 11–13 are based on the following information and illustration.

A 500 MW$_e$ plant operates with the following power history for a period of one year. Three month periods are shown on the x-axis as t_1, t_2, and so on.

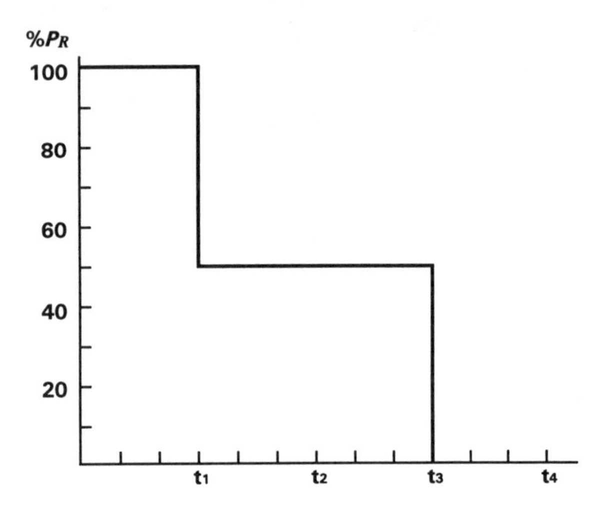

In the three month period from t_3 to t_4 the plant was shutdown for scheduled maintenance and upgrades.

NUCLEAR FUEL MANAGEMENT–11

What is the capacity factor?

(A) 0.25
(B) 0.50
(C) 0.67
(D) 0.75

Solution:

$$\text{CF} = \int_0^T \frac{P(t)}{P_R T}\,dt$$

P_R = rated electric power
t_1,t_2,t_3,t_4 = quarter periods [three month periods]
T = time period [usually one year]

Split the integral into quarters, note the power during any given quarter is constant and thus can be pulled out of the integral, and substitute given information.

$$\text{CF} = \int_0^T \frac{P(t)}{P_R T}\,dt$$

$$= \left(\frac{1}{P_R T}\right)\left(P_{0,1}\int_{t_0}^{t_1} dt + P_{1,3}\int_{t_1}^{t_3} dt + P_{3,4}\int_{t_3}^{t_4} dt\right)$$

$$= \left(\frac{1}{(500\text{ MW})(1\text{ yr})}\right)\Big((500\text{ MW})(0.25\text{ yr}) + (0.50)(500\text{ MW})(0.5\text{ yr}) + (0\text{ MW})(0.25\text{ yr})\Big)$$

$$= \boxed{0.50\ \ (50\%)}$$

Answer is B.

NUCLEAR FUEL MANAGEMENT–12

What is the availability factor for the time period shown?

(A) 0.25
(B) 0.50
(C) 0.75
(D) 1.00

Solution:

The *availability factor* AF is defined as follows[29]

$$\text{AF} = \frac{\left(\begin{array}{c}\text{time during which the plant}\\ \text{was operational in time period } T\end{array}\right)}{T}$$

$$= \frac{9\text{ months}}{12\text{ months}}$$

$$= \boxed{0.75\ \ (75\%)}$$

Answer is C.

[29] The AF > CF. The availability factor is a measure of the time the plant was available and the capacity factor is a measure of how close to rated power a plant operates during this time.

NUCLEAR FUEL MANAGEMENT–13

What is most nearly the number of effective full power days (EFPD) the plant experienced during this time period?

(A) 90 EFPD
(B) 180 EFPD
(C) 270 EFPD
(D) 360 EFPD

Solution:

The *effective full power day* (EFPD) is defined as

$$\text{EFPD} = \int_0^T \text{CF}(t)dt$$

$\text{CF}(t)$ = capacity factor as a function of time
T = time period in days

This can be solved a couple of different ways. First, since we know $\text{CF}(t) = 0.50$ for the year, $\text{CF}(t)$ is a constant and

$$\begin{aligned}\text{EFPD} &= \int_0^T \text{CF}(t)dt \\ &= \text{CF}\int_0^T dt \\ &= (0.50)\left(t\Big|_0^{365}\right) = (0.50)(365\text{ days} - 0\text{ days}) \\ &= \boxed{182.5\text{ days}\quad(180\text{ days})}\end{aligned}$$

Alternately, one can perform the integration over each time period giving[30]

$$\begin{aligned}\text{EFPD} &= \int_0^T \text{CF}(t)dt \\ &= \text{CF}_{0,1}\int_{t_0}^{t_1} dt + \text{CF}_{1,2}\int_{t_1}^{t_2} dt + \text{CF}_{2,3}\int_{t_2}^{t_3} dt + \text{CF}_{3,4}\int_{t_3}^{t_4} dt \\ &= (1.0)(91.5\text{ days}) + (0.5)(91.25\text{ days}) \\ &\quad + (0.5)(91.25\text{ days}) + (0)(91.25\text{ days}) \\ &= \boxed{182.5\text{ days}\quad(180\text{ days})}\end{aligned}$$

Answer is B.[31]

NUCLEAR FUEL MANAGEMENT–14

A certain PWR is a three-batch core. The *excess reactivity* in the core $\rho_N(t)$ is governed by the following expression[32]

$$\rho_N(t) = \sum_{n-1}^{N} \frac{\left(1 - \text{BU}_N^n(t)\right)(\rho_1)}{N}$$

N = number of batches in the core
n = cycle number
ρ_1 = excess reactivity in a one batch core

The final burnup is governed by

$$\text{BU}_N(T) = \left(\frac{2N}{N+1}\right)\left(\text{BU}_1(t)\right)$$

T = time period of concern
BU_N = the final burnup of a batch in a N batch core, once that batch has been in the core for N cycles[33]

For the same initial enrichment, increasing the number of batches in a core does which of the following:

(A) decrease the time between refuelings
(B) increase the availability factor
(C) increase the final burnup by one and a half times over a single batch core
(D) A and C above

[30] During any three month period, the actual number of days used can vary depending on the accounting method (e.g., 90 days is common). For EFPDs and other terms related to energy usage, the exact number of days is important in order to determine fuel consumption, decay heat generation, and so on.

[31] Note that EFPD multiplied by P_R equals the energy expended in megawatt days (MWD). Recall the P_R is the rated power.

[32] The equation assumes a linear relationship between reactivity and burnup. Additionally, the assumption is made that the total core reactivity, at any time in a cycle, is equal to the average of the sums of the individual reactivities.

[33] The units are MWD/MTU or similar units.

Solution:

Consider each item in turn.

As the number of batches increases, the number of refuelings also increases in order to maintain an equilibrium cycle. (An equilibrium cycle begins after the Nth batch. At this point all subsequent reloads are identical as are the fuel assembly loading patterns.) Thus, increasing the number of batches decreases the time between refuelings and A is an answer.

The availability factor AF as defined earlier depends on the time the plant is available for producing power. By increasing the number of refuelings, the total shutdown time is longer and the availability drops. Therefore, B is not an answer.

Comparing burnup for a three-batch core with the burnup for a single-batch core gives

$$\mathrm{BU}_N(T) = \left(\frac{2N}{N+1}\right)\left(\mathrm{BU}_1(t)\right)$$

$$\begin{aligned}\mathrm{BU}_3(T) &= \left(\frac{(2)(3)}{3+1}\right)\left(\mathrm{BU}_1(t)\right)\\ &= \left(\frac{6}{4}\right)\left(\mathrm{BU}_1(t)\right)\\ &= \tfrac{3}{2}\,\mathrm{BU}_1(t)\end{aligned}$$

Therefore, C is an answer.[34] This makes A and C an acceptable answer.

Answer is D.

Note: The reactivity swings also decrease as the number of batches increase. The lower the reactivity swing, the less control of reactivity is required (i.e., the fewer poisons or control rods are necessary). With ρ as the excess reactivity plotted against the burnup for single, two-batch, and three-batch core, the reactivity swing can be represented graphically as shown.

[34] In an attempt to improve the burnup (i.e., increase it), the number of batches can be increased. In the extreme, with an infinite number of batches, the burnup increases by a factor of two over $\mathrm{BU}_1(T)$. (This can be seen by taking the limit of the burnup equation as N approaches infinity.) The downside to increasing the number of batches is that plant shutdown time for refuelings increases. A economic balance must be struck.

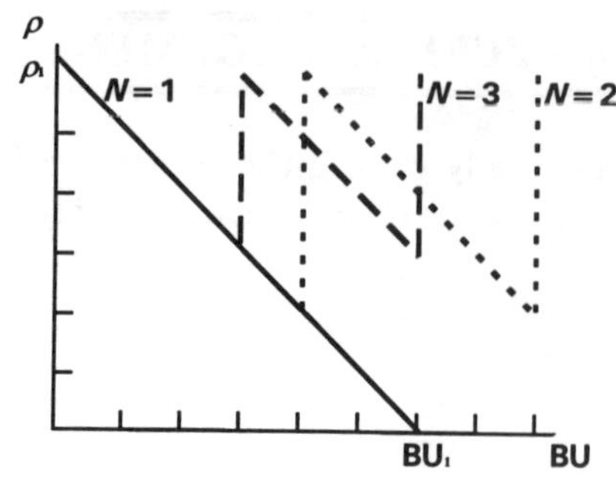

The reactivity swing, referenced to the beginning of the cycle is

$$\rho_2(0) = 2/3\left|\rho_1\right|$$

$$\rho_3(0) = 1/2\left|\rho_1\right|$$

Nuclear Fuel Management (NFM) problems 15 and 16 are based on the following information and illustration.

A 1000 MW$_e$ PWR operates at full power and 25% efficiency for 11 months. The decay heat removal (DHR) requirements must be determined to ensure adequate cooling. One method of decay heat generation determination is to utilize the following formula (or one similar).[35]

$$\frac{P(t_o, t_s)}{P_o} = \frac{P(t_s)}{P_o} - \frac{P(t_o + t_s)}{P_o}$$

P = power generated
P_o = thermal power
t_o = operating time in seconds
t_s = time since shutdown in seconds

As decay heat generation is directly related to the number of fissions and thus the thermal power, a normalized graph, this one adapted from the standard ANS-5, can be generated as shown.

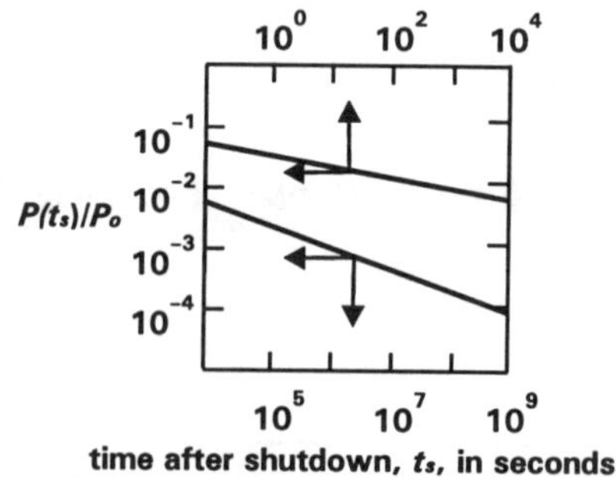

[35] Several formulas exist. Some use coefficients for each term whose value depends on plant conditions and operating history. The formula used depends on regulatory requirements and the accuracy desired.

NUCLEAR FUEL MANAGEMENT–15

What is the decay heat generation power removal requirement 24 hours after shutdown?

(A) 4 MW
(B) 17 MW
(C) 70 MW
(D) 280 MW

Solution:

The decay heat power can be determined from the given formula and the graph. First, determine the entering arguments for the graph.

The operating time for this reactor in seconds (t_o) is[36]

$$t_o = (11\,\text{months})\left(\frac{30\,\text{days}}{1\,\text{month}}\right)\left(\frac{24\,\text{hr}}{1\,\text{day}}\right)\left(\frac{60\,\text{min}}{1\,\text{hr}}\right)\left(\frac{60\,\text{s}}{1\,\text{min}}\right)$$
$$= 2.85\times 10^7\ \text{s}$$

The time since shutdown in seconds (t_s) is

$$t_s = (24\,\text{hr})\left(\frac{60\,\text{min}}{1\,\text{hr}}\right)\left(\frac{60\,\text{s}}{1\,\text{min}}\right)$$
$$= 8.64\times 10^4\ \text{s}$$

The combined operating and shutdown time (which is a necessary number in the decay heat generation formula) is

$$t_o + t_s = 2.85\times 10^7\ \text{s} + 8.64\times 10^4\ \text{s}$$
$$= 2.86\times 10^7\ \text{s}$$

The thermal power is[37]

$$P_o = \frac{P_{\text{out}}}{\eta_{\text{th}}} = \frac{1000\ \text{MW}}{0.25} = 4000\ \text{MW}$$

P_{out} = electrical power output
η_{th} = thermal efficiency

[36] The calculation used 30 days per month. To more exactly determine the decay heat generation one would need to know the actual number of days.
[37] The thermal power, not the electrical power, is the correct power to use because decay heat generation depends on the number of fissions occurring, including those whose energy is lost as waste heat.

Substituting into the formula gives

$$\frac{P(t_o, t_s)}{P_o} = \frac{P(t_s)}{P_o} - \frac{P(t_o + t_s)}{P_o}$$

$$\frac{P(2.85\times 10^7\ \text{s},\ 8.64\times 10^4\ \text{s})}{P_o} = \frac{P(8.64\times 10^4\ \text{s})}{P_o} - \frac{P(2.86\times 10^7\ \text{s})}{P_o}$$

From the graph,

$$\frac{P(8.64\times 10^4\ \text{s})}{P_o} \approx 5\times 10^{-3}$$

$$\frac{P(2.86\times 10^7\ \text{s})}{P_o} \approx 8\times 10^{-4}$$

Substituting,

$$\frac{P(2.85\times 10^7\ \text{s},\ 8.64\times 10^4\ \text{s})}{P_o} = 5\times 10^{-3} - 8\times 10^{-4}$$
$$= 4.20\times 10^{-3}$$

Rearranging the equation to solve for the decay heat power that is generated 24 hours after shutdown gives

$$P(2.85\times 10^7\ \text{s},\ 8.64\times 10^4\ \text{s}) = P_o(4.20\times 10^{-3})$$
$$= (4000\ \text{MW})(4.20\times 10^{-3})$$
$$= \boxed{16.8\ \text{MW}\quad (17\ \text{MW})}$$

Answer is B.

NUCLEAR FUEL MANAGEMENT–16

By approximately what factor is the decay heat generation requirement greater immediately upon shutdown as compared to the 24 hour period calculated in the Prob. 15?

(A) 2
(B) 10
(C) 15
(D) 100

Solution:

The operating time remains the same (t_o = 2.85 × 10^7 s) and the output power remains the same (4000 MW). The time shutdown is zero, or to use the graph, 10^{-1} s (the shortest time available on the graph). Therefore,

$$\begin{aligned} t_o + t_s &= 2.85\times10^7 \text{ s} + 10^{-1} \text{ s} \\ &= 2.85\times10^7 \text{ s} \end{aligned}$$

The decay heat power can be determined from

$$\frac{P(t_o,t_s)}{P_o} = \frac{P(t_s)}{P_o} - \frac{P(t_o+t_s)}{P_o}$$

$$\frac{P(2.85\times10^7 \text{ s, } 10^{-1} \text{ s})}{P_o} = \frac{P(10^{-1})}{P_o} - \frac{P(2.85\times10^7)}{P_o}$$

Determining the values of the terms on the right side of the equation from the graph gives

$$\begin{aligned} \frac{P(2.85\times10^7 \text{ s, } 10^{-1} \text{ s})}{P_o} &\cong 7\times10^{-2} - 8\times10^{-4} \\ &\cong 6.92\times10^{-2} \\ P(2.85\times10^7 \text{ s, } 10^{-1} \text{ s}) &= (6.92\times10^{-2})(P_o) \\ &= (6.92\times10^{-2})(4000 \text{ MW}) \\ &= 2.77\times10^2 \text{ MW} \quad (276.8 \text{ MW}) \end{aligned}$$

Thus, the decay heat generation immediately upon shutdown[38] divided by the decay heat generation at the 24 hour point results in the factor desired.

$$\frac{276.8 \text{ MW}}{16.8 \text{ MW}} = \boxed{16.5 \quad (15)}$$

Answer is C .

[38] This amount of decay heat is approximately 7% of the rated thermal power output. At the 24 hour point the decay heat is <1% of the rated thermal power output. Both these percentages are good approximations for decay heat generation regardless of the plant.

Nuclear Fuel Management (NFM) problems 17–20 are based on the following information and tables and represents a partial excerpt from 10CFR61.[39]

activity concentrations used to classify low-level wastes (LLW)

Table A

nuclide	concentration (Ci/m³)
Ni-59	220
Tc-99	3
I-129	0.08
alpha-emitting transuranic waste (TRU) with a $T_{1/2}$ >5 yr	100 nCi/g
Pu-241	3500 nCi/g

Table B

	concentration (Ci/m³)		
nuclide	column 1	column 2	column 3
$T_{1/2}$ < 5 yr	700	*	*
^3H	40	*	*
Ni-63	3.5	70	700
Sr-90	0.04	150	7000
Cs-137	1	44	4600

*No specific limit is established. Heat generation or external radiation may limit their concentration. These wastes are class B unless other considerations make them class C.

Classification guidance using the above tables can be found in Table 2.3 of App. 2.B.

[39] Title 10 of the Code of Federal Regulations, Chapter 61. Consult the latest revision to determine up-to-date requirements.

NUCLEAR FUEL MANAGEMENT–17

A LLW package is determined to contain Ni-59 at 0.02 mCi/cm^3 and Ni-63 at 1 Ci/cm^3. What is the classification of this waste?

(A) class A
(B) class B
(C) class C
(D) GTCC

Solution:

First, convert Ni-59 concentration units to those used in Table A.

$$[^{59}\mathrm{Ni}] = \left(0.02\,\frac{\mathrm{mCi}}{\mathrm{cm}^3}\right)\left(\frac{1\,\mathrm{Ci}}{10^3\,\mathrm{mCi}}\right)\left(\frac{10^2\,\mathrm{cm}}{1\,\mathrm{m}}\right)^3 = 20\,\mathrm{Ci/m^3}$$

This concentration of Ni-59 is less than 0.1 times the concentration listed in Table A (i.e., less than 22 Ci/m^3), therefore, the waste is consider class A.

The Ni-63 concentration is less than column one of Table B and is therefore also a class A waste.

Thus, the package is class A.

Answer is A .[40]

NUCLEAR FUEL MANAGEMENT–18

A LLW package contains Sr-90 in a concentration of 75 Ci/m^3 and Cs-137 in a concentration of 25 μCi/cm^3. What is the classification of this package?

(A) class A
(B) class B
(C) class C
(D) GTCC

Solution:

Change the units of the Cs-137 to those used in Table B.

$$[^{137}\mathrm{Cs}] = \left(25\,\frac{\mu\mathrm{Ci}}{\mathrm{cm}^3}\right)\left(\frac{1\,\mathrm{Ci}}{10^6\,\mu\mathrm{Ci}}\right)\left(\frac{10^2\,\mathrm{cm}}{\mathrm{m}}\right)^3 = 25\,\mathrm{Ci/m^3}$$

The concentration of the Cs-137 exceeds the value listed in column one of Table B but is less than the value in column two; therefore, this portion of the waste is class B.

The concentration of Sr-90 exceeds the value in column one of Table B and is also less than the value in column two; therefore, this portion of the waste is class B.

As the two wastes are packaged together, the sum of the fractions must be used to determine the classification as specified in table footnote "b" of Table 2.3 in App. 2.B.

For the Cs-137, comparing the concentration to the column two limit gives

$$\frac{\text{concentration}}{\text{limit}} = \frac{25\,\frac{\mathrm{Ci}}{\mathrm{m}^3}}{44\,\frac{\mathrm{Ci}}{\mathrm{m}^3}} = 0.57$$

For the Sr-90, comparing the concentration to the column two limit gives

$$\frac{\text{concentration}}{\text{limit}} = \frac{75\,\frac{\mathrm{Ci}}{\mathrm{m}^3}}{150\,\frac{\mathrm{Ci}}{\mathrm{m}^3}} = 0.50$$

Thus, the sum of the fractions is

$$0.57 + 0.50 = 1.07 > 1.0$$

Thus, the waste can not be classified using column two limits. A repeat of the above procedure, or a quick mental calculation using column three limits, indicates the sum of the fractions is easily less than 1.0.

Therefore, the value of the activity concentration is between columns two and three in Table B making the package class C waste .

Answer is C.

[40] The sum of fractions method is not applicable here as this is a mixture of radionuclides from two separate tables. Therefore, the most restrictive classification applies.

NUCLEAR FUEL MANAGEMENT–19

A LLW package contains Tc-99 at a concentration of 0.02 Ci/m^3 and tritium at a concentration of 45 Ci/m^3. What is the classification of this waste?

(A) class A
(B) class B
(C) class C
(D) GTCC

Solution:

The Tc-99 concentration of 0.02 Ci/m^3 is less than 0.1 of the limit of 3 Ci/m^3 listed in Table A. Therefore, this portion of the waste is class A.

The H-3 concentration of 45 Ci/m^3 exceeds the column one limit of Table B. The asterisk in column two indicates that unless other considerations are applicable, this is a class B waste.

Since the most restrictive classification from the two tables applies in accordance with table footnote "a" of Table 2.3 in App. 2.B, the package is class B.

Answer is B.

NUCLEAR FUEL MANAGEMENT–20

A LLW package with dimensions of 12 in × 12 in × 8 in contains 20 000 μCi of ^{32}P. What is the classification of this package?

(A) class A
(B) class B
(C) class C
(D) GTCC

Solution:

The P-32 is not specifically listed in Table A or B. It does however fall into the Table B category "$T_{1/2} < 5$ yr." From Table 2.4 in App. 2B, the half-life is approximately 14 days.

The activity is given; to determine the concentration first determine the volume of the package.

$$(12\ \text{in})(12\ \text{in})(8\ \text{in}) = 1152\ \text{in}^3$$

Convert this value to SI units.[41]

$$\left(1152\ \text{in}^3\right)\left(\frac{2.54\ \text{cm}}{1\ \text{in}}\right)^3\left(\frac{1\ \text{m}}{10^2\ \text{cm}}\right)^3 = 0.02\ \text{m}^3$$

The concentration is therefore

$$\begin{aligned}\left[{}^{32}\text{P}\right] &= \frac{20\,000\times 10^{-6}\ \text{Ci}}{0.02\ \text{m}^3}\\ &= 1.00\ \text{Ci/m}^3\end{aligned}$$

This concentration is less than the limit in column 1 of Table B (700 Ci/m^3). Therefore, this package is class A.

Answer is A.

Nuclear Fuel Management (NFM) problems 21 and 22 are based on the following information and equation.

There are five general categories of radioactive wastes:

High-Level Wastes (HLW). Defined in the Nuclear Waste Policy Act as the "highly radioactive material from reprocessing spent fuel…."[42]
Transuranic Wastes (TRU). Defined as alpha emitting isotopes with an atomic number greater than 92, half-lives greater than five years and activity concentrations greater than 100 nCi/g of waste.[43]
Low-Level Waste (LLW). Defined as material that is not HLW, spent nuclear fuel, TRU, or by-product material.[44]
Uranium mill tailings. These are treated differently from other radioactive wastes since they are not transported from their generation point.
Naturally Occurring and Accelerator produced Radioactive Materials (NARM). These wastes are not regulated by the Nuclear Regulatory Commission (NRC) but by individual states and generally classified as LLW.

---

[41] Package volume reduction can lower costs, but as shown here, one must be cognizant of the fact that decreasing the volume results in an increase in the concentration and may change the classification of the waste. Standard volume reduction methods include compaction, evaporation, and incineration.
[42] There are three general groups of radioisotopes in spent fuel: fission products, *actinides*, and activation products.
[43] This is actually a subset of HLW. Further, TRUs have been defined as a category only in the U.S.
[44] Questions in the Nuclear Fuel Management section more clearly define low level wastes. See Table 2.3 in App. 2.B for additional information.

Radioactive wastes generate heat. Spent fuel is often stored in a pool upon initial removal from a reactor in order to remove this heat.[45] One method of determining decay heat generation is using the equation found in an American Nuclear Society standard 5.1 for decay heat generation calculation. Specifically,

$$\frac{P(t,\infty)}{P_o} = \mathrm{A}t^{-\mathrm{a}}$$

$P(t,\infty)$ = decay power at t seconds after shutdown, following operation at P_o for an infinite time
t = time in seconds following reactor shutdown
A,a = constants

constants "A" and "a" of ANS 5.1 Standard

time interval (sec)	A	a
$0.1 < t < 10$	0.0603	0.0639
$10 < t < 150$	0.0766	0.181
$150 < t < 4 \times 10^6$	0.130	0.283
$4 \times 10^6 < t < 2 \times 10^8$	0.266	0.335

NUCLEAR FUEL MANAGEMENT–21

An 1150 $\mathrm{MW_e}$ PWR with a plant efficiency of 30% has been shutdown for 30 days after operating for an extended time at rated power. If the spent fuel is to be moved to a pool, what is the heat removal capacity required?

(A) 1.2×10^2 Btu/hr
(B) 2.1×10^2 Btu/hr
(C) 7.6×10^6 Btu/hr
(D) 2.6×10^7 Btu/hr

Solution:

Since the plant operated for an extended time, the infinity assumption in ANS 5.1 Standard is valid and the following formula can be used.

$$\frac{P(t,\infty)}{P_o} = \mathrm{A}t^{-\mathrm{a}}$$

The time since shutdown in seconds is

$$\begin{aligned} t &= (30\ \text{days})\left(\frac{24\ \text{hr}}{1\ \text{day}}\right)\left(\frac{60\ \text{min}}{1\ \text{hr}}\right)\left(\frac{60\ \text{sec}}{1\ \text{min}}\right) \\ &= 2.59 \times 10^6\ \text{sec} \end{aligned}$$

Using the table for the calculated time gives A = 0.130 and a = 0.283. The term P_o is the constant power preceding the shutdown—the rated power (P_R) in this case. Thus,

$$\begin{aligned} P_R &= \frac{P_{\text{out}}}{\eta} \\ &= \frac{1150\ \text{MW}}{0.30} \\ &= 3833.3\ \text{MW} \end{aligned}$$

Rearranging and substituting into the decay power equation gives

$$\begin{aligned} \frac{P(t,\infty)}{P_o} &= \mathrm{A}t^{-\mathrm{a}} \\ &= \mathrm{A}t^{-\mathrm{a}} P_o \\ &= (0.130)(2.59 \times 10^6\ \text{sec})^{-0.283}(3833.3\ \text{MW}) \\ &= 7.63\ \text{MW} \end{aligned}$$

Converting to the requested units gives

$$\begin{aligned} P(t,\infty) &= 7.63\ \text{MW} \\ &= (7.63\ \text{MW})\left(\frac{10^6\ \text{W}}{1\ \text{MW}}\right)\left(\frac{3.413\,\frac{\text{Btu}}{\text{hr}}}{1\ \text{W}}\right) \\ &= \boxed{2.6 \times 10^7\ \text{Btu/hr}} \end{aligned}$$

Answer is D.

[45] Dry cask storage is also used but only after the decay heat generation is reduced to a rate where the heat removal capability of the cask is capable of maintaining the required temperature.

NUCLEAR FUEL MANAGEMENT–22

Approximately 4 mCi of Pu-240 is to be disposed of in a package containing 1.1 kg of material with essentially negligible activity. What is the classification of this waste?

(A) LLW class A
(B) LLW class C
(C) TRU
(D) either B or C

Solution:

Since the Pu-240 concentration limit (i.e., as a alpha emitting TRU with a half-life greater than five years) in Table A of the information section for NFM Probs. 17–20 is in terms of nCi/g, first determine the concentration for the plutonium in these terms.

$$\left[{}^{240}\text{Pu}\right] = \frac{(4\text{ mCi})\left(\dfrac{1\text{ Ci}}{10^3\text{ mCi}}\right)}{(1.1\text{ kg})\left(\dfrac{10^3\text{g}}{1\text{kg}}\right)}$$
$$= \left(3.64\times10^{-6}\ \frac{\text{Ci}}{\text{g}}\right)\left(\frac{10^9\text{ nCi}}{1\text{Ci}}\right)$$
$$= 3.64\times10^3\text{ nCi/g}\quad(3636\text{ nCi/g})$$

This concentration is greater than 0.1 the value in Table A of NFM Probs. 17–20 and thus the waste is not a LLW class A. Further, the concentration exceeds the 100 nCi/g limit given in Table A and thus it is not LLW class C either, although it would fall into GTCC category according to the guidance in Table 2.3 of App. 2.B. However, this is not a choice.

The atomic number of Pu-240 is 94 (>92); its half-life is over 6000 yr according to the information in Table 2.4 of App. 2.B; and it is present in a concentration greater than 100 nCi/g. Therefore, this waste meets all the requirements to be classified as a transuranic waste (TRU).

Answer is C.

NUCLEAR FUEL MANAGEMENT–23

Current plans call for the disposal of HLW in a deep geological repository. This disposal method will also be used for TRU, for which there is no essential difference with HLW regarding treatment and disposal. The largest portion of the HLW and TRU is in liquid form. Disposal is expected to occur in solid form. Possible methods follow.

α	calcination
β	cementation
γ	vitrification

Partial descriptions of the methods follow.

1	mix with cement; pour into a container and allow to dry
2	mix with glass frit
3	heat until completely dry

Match the methods given with their descriptions. Also, specify the preferred method.

(A) α1, β2, γ3/calcination
(B) α2, β1, γ3/vitrification
(C) α2, β3, γ1/cementation
(D) α3, β1, γ2/vitrification

Solution:

Calcination results when a liquid is heated until it is a completely dry powder-like substance (α3).

Cementation is the process of mixing the waste with cement, pouring the mix into a mold and allowing it to dry into a concrete block (β1).

Vitrification is mixing the waste with glass frit to form a solid glass (γ2).

The preferred method is vitrification due to its low leach rate and low solubility in water; high stability for HLW and TRU nuclides; resistance to radiation damage as well as significant evidence of its stability over thousands of years.[46] Thus, the answer is α3, β1, γ2/vitrification.

Answer is D .

[46] As evidenced by ancient artifacts made of similar material.

Nuclear Fuel Management (NFM) problems 24 and 25 are based on the following information and equation.

Transportation of radioactive materials is governed by the U.S. Department of Transportation (DOT);[47] the Nuclear Regulatory Commission (NRC);[48] and the U.S. Postal Service.[49] International transportation is further governed by International Atomic Energy Agency (IAEA) regulations.

Differentiation between HLW and LLW does not occur. Instead, packaging requirements depend on the isotopes in a given package and the activity associated with those isotopes. Numerous definitions of material and classifications exist.

One such classification is low specific activity (LSA) material. A partial definition of LSA material follows.

LSA = material which has essentially uniformly distributed radioactivity in which the average concentration per gram does not exceed the requirements of 10CFR71

A partial table from 10CFR71 for selective isotopes follows.

isotope	A1 special form (Ci)	A2 normal form (Ci)
Co-60	7	7
Cs-137	30	10
Pb-210	100	0.2
Ra-226	10	0.05
U-233	100	0.1

A2 = the maximum activity of radioactive material, other than special form radioactive material, permitted in a type A package

A material is considered LSA material if it meets the following requirements from 10CFR71 regarding the average concentration per gram of activity:

(i) no more than 0.0001 mCi from radionuclides with A2 activity ≤ 0.05 Ci

(ii) no more than 0.005 mCi from radionuclides with A2 activity more than 0.05 Ci but less than 1 Ci

(iii) no more than 0.3 mCi from radionuclides with A2 activity more than 1 Ci

If a mixture of radionuclides is to be classified as LSA, it must meet the following condition (know as the ratio rule).

$$\frac{A}{0.0001}+\frac{B}{0.005}+\frac{C}{0.3}<1.0$$

A = the total activity in mCi/g of all nuclides with an A2 activity ≤ 0.05 Ci

B = the total activity in mCi/g of all nuclides with an A2 activity > 0.05 Ci but < 1.0 Ci

C = the total activity in mCi/g of all nuclides with an A2 activity ≥ 1.0 Ci

Consider the following radioactive material for transport.

	material	estimated concentration per gram
I	Ra-226	0.1 μCi
II	Pb-210	50 μCi
III	U-233	500 pCi
IV	Cs-137	3500 μCi

NUCLEAR FUEL MANAGEMENT–24

If the material is shipped separately, which packages can be shipped as LSA material in a type A package?

(A) I and II
(B) I and III
(C) II and IV
(D) I, II, and III

Solution:

To be designated as LSA material and be shipped in a type A package, the requirements listed in the information section excerpted from 10CFR71 must be met. As the material is to be shipped in separate packages, only conditions (i), (ii), and (iii) must be met; the ratio rule is not necessary. To compare the requirements with the actual activities, first calculate the activities in terms of the units used in the regulations.

[47] See 49CFR100–199.
[48] See 10CFR71.
[49] See 39CFR124.

The activity of the Ra-226 (α_{Ra}) is

$$\alpha_{Ra} = (0.1\ \mu\text{Ci})\left(\frac{1\ \text{Ci}}{10^6\ \mu\text{Ci}}\right)\left(\frac{10^3\ \text{mCi}}{1\ \text{Ci}}\right) = 0.0001\ \text{mCi}$$

This activity equals the maximum allowed by condition (i). Thus, Ra-226 in this instance is LSA material if shipped separately. Item I is an answer.

The A2 activity of Pb-210 as given in the excerpt from 10CFR71 is 0.2 Ci which is between 0.05 Ci and 1.0 Ci. Therefore, the maximum allowed activity is 0.005 mCi as indicated by condition (ii). The actual activity is

$$\alpha_{Pb} = (50\ \mu\text{Ci})\left(\frac{1\ \text{Ci}}{10^6\ \mu\text{Ci}}\right)\left(\frac{10^3\ \text{mCi}}{1\ \text{Ci}}\right) = 0.050\ \text{mCi}$$

This activity level is greater than the 0.005 mCi limit of condition (ii) and thus the material is not the LSA type. Item II is not an answer.

The U-233 A2 activity is 0.1 Ci which is also between 0.05 Ci and 1.0 Ci. Again the maximum activity is 0.005 mCi from condition (ii). The actual activity is

$$\alpha_{U} = (500\ \text{pCi})\left(\frac{1\ \text{Ci}}{10^{12}\ \text{pCi}}\right)\left(\frac{10^3\ \text{mCi}}{1\ \text{Ci}}\right) = 0.005\times 10^{-3}\ \text{mCi}$$

This activity level is less than the 0.005 mCi limit of condition (ii) and thus the material is LSA type if shipped separately. Item III is an answer.

The Cs-137 A2 activity is 10 Ci which is greater than 1.0 Ci. Thus the maximum activity is 0.3 mCi from condition (iii). The actual activity is

$$\alpha_{CS} = (3500\ \mu\text{Ci})\left(\frac{1\ \text{Ci}}{10^6\ \mu\text{Ci}}\right)\left(\frac{10^3\ \text{mCi}}{1\ \text{Ci}}\right) = 3.5\ \text{mCi}$$

This activity level is greater than the 0.3 mCi limit of condition (iii) and thus the material is not the LSA type.

Therefore, items I and III are LSA material if shipped separately.

Answer is B.

NUCLEAR FUEL MANAGEMENT–25

Which combination of materials can be transported together and retain the LSA designation?

(A) no combination
(B) I and II
(C) I and III
(D) II and III

Solution:

Since a mixture of radionuclides is to be shipped together, the ratio rule must be used.

$$\frac{A}{0.0001} + \frac{B}{0.005} + \frac{C}{0.3} < 1.0$$

Recall that the activities represented by A are those with A2 activities less than or equal to 0.05 Ci (Ra-226); for B between 0.05 Ci and 1.0 Ci (Pb-210 and U-233); for C greater than 1.0 Ci (Cs-137). Importantly, all the entering arguments for A, B, and C are to be in mCi/g. Keeping this in mind, the following terms in the ratio rule equation can be calculated.

$$\begin{aligned}\frac{A}{0.0001} &= \frac{[\text{Ra-226}]}{0.0001} \\ &= \frac{0.0001\ \frac{\text{mCi}}{\text{g}}}{0.0001} \\ &= 1\ \text{mCi/g}\end{aligned}$$

$$\begin{aligned}\frac{B}{0.005} &= \frac{[\text{Pb-210}]}{0.005} \\ &= \frac{0.050\ \frac{\text{mCi}}{\text{g}}}{0.005} \\ &= 10\ \text{mCi/g}\end{aligned}$$

$$\begin{aligned}\frac{B}{0.005} &= \frac{[\text{U-233}]}{0.005} \\ &= \frac{0.005\times 10^{-3}\ \frac{\text{mCi}}{\text{g}}}{0.005} \\ &= 1\times 10^{-3}\ \text{mCi/g}\end{aligned}$$

$$\frac{C}{0.3} = \frac{[\text{Cs-137}]}{0.3} = \frac{3.5 \frac{\text{mCi}}{\text{g}}}{0.3} = 11.7 \text{ mCi/g}$$

Note: These terms are normally shown as unitless given that the entering arguments are expected to be in mCi/g.

Place the information just calculated into a table for ease of use.

material	A/0.0001	B/0.005	C/0.3
Ra-226	1		
Pb-210		10	
U-233		0.001	
Cs-137			11.7

Combine each of the terms per the ratio rule.

$$\frac{A}{0.0001} + \frac{B}{0.005} + \frac{C}{0.3} < 1.0$$

The result is that no combination of materials results in a value less than one. Thus, none of the material can be shipped together and retain the LSA designation.

Answer is A.

9 NUCLEAR RADIATION: SOLUTIONS

Nuclear Radiation (NR) problems 1–3 are based on the following information and illustration. [1]

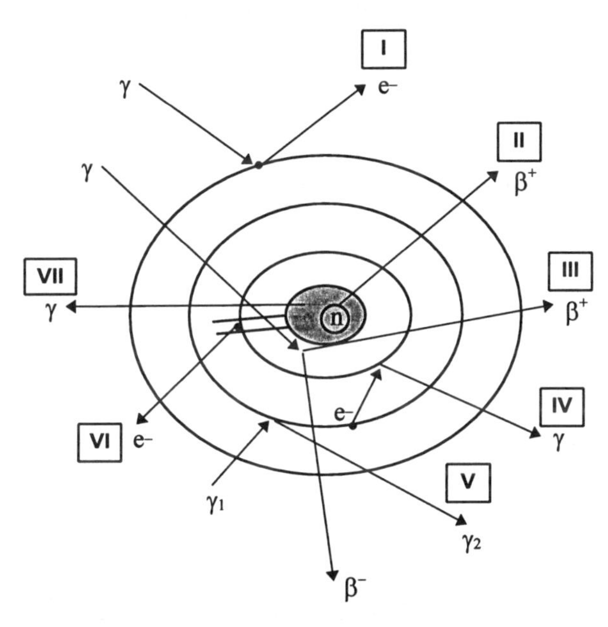

γ = photons [subscripts indicate different frequencies]
e– = electrons
β = beta particles [electrons/positrons]

NUCLEAR RADIATION–1

Which process is referred to as the photoelectric effect?

(A) I
(B) II
(C) IV
(D) VI

[1] The orbits of the electrons are shown as circular. This is certainly not strictly true. Indeed, one should speak of the orbits in terms of position probability. While not required for the nuclear professional engineer examination, an understanding of the atoms at this level means having an understanding of quantum theory. See the Reference section for more information.

Solution:

Consider each item in reverse order.

Process VI represents *internal conversion.* An excited nucleus can interact with one of its innermost electrons and in doing so transfer excitation energy to this electron which then manifests itself as kinetic energy. The electron is ejected from the atom with an energy equal to the excitation energy (i.e., the nuclear transition energy) of the nucleus minus the ionization energy of that particular electron. The answer is not D.

When the ejected electron is no longer in its original orbit, an outer electron can move into this lower energy orbit and emit an *x*-ray to rid itself of the excess energy.[2] This is represented by process IV. Therefore, C is not the answer.

Process II represents the β^+ (positron) decay of a proton.[3] This tends to occur in nuclides where too few neutrons exist for stability.[4] Therefore, B is not the answer.

Process I shows an incoming photon interacting with an electron, disappearing, and ejecting the electron from control of the atom.[5] The kinetic energy of the electron equals the energy of the incoming photon minus the binding energy of the electron from the atom. The electron is ejected at right angles to the incoming photon, indicating a low-energy photon. (High-energy photons cause the electron to be ejected more along the same path as the incoming particle.) This entire process is called the photoelectric effect.

[2] A photon emitted due to electron transitions or by stopping high-velocity electrons is called a *x*-ray. A high-energy photon emitted during a nuclear transition is called a gamma ray. In spite of these definitions, the Greek letter gamma (γ) is often used to represent photons of both types. The *x*-rays and gamma rays, except for their energy differences, differ only in their sources (i.e., electron movement for *x*-rays and the nucleus for gamma rays).

[3] In this context, these and the β^- are sometimes referred to as a beta rays.

[4] See the Nuclear Theory section for more information.

[5] The interaction usually occurs with the entire atom.

The probability of the interactions is

$$P_{PEE} \propto \frac{Z^n}{E^3}$$

Z = atomic number
n = exponent which depends on the photon energy (3 for low energy up to 5 for high energy)
E = energy of the photon

Answer is A.

NUCLEAR RADIATION–2

Which process is referred to as Compton scattering?

(A) I
(B) IV
(C) V
(D) VI

Solution:

Process I, IV, and VI were explained in Prob. 1.

Process V represents Compton scattering and is sometimes called elastic scattering as both energy and momentum are conserved. In this process the incoming photon is scattered through interaction with an orbital electron, usually a loosely bound electron, transferring some of its energy to the electron and emerging with a lower energy and thus a longer wavelength. The wavelengths are related by

$$\lambda_1 - \lambda_2 = \lambda_c(1 - \cos\phi)$$

$$\lambda_c = \frac{h}{m_e c} = 2.426 \times 10^{-10} \text{ cm}$$

λ_c = Compton wavelength
h = Planck's constant
m_e = rest mass of an electron
c = speed of light

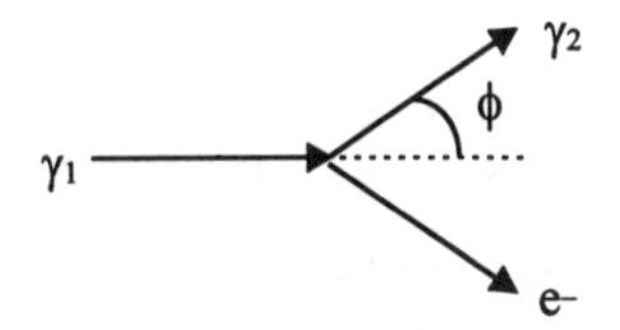

The probability of such interactions is[6]

$$P_{cs} \propto \frac{Z}{E}$$

Z = atomic number
E = energy of the photon

Note: The important aspect of this problem, with regard to shielding, is that this interaction makes shielding calculations more difficult since the photon does not disappear as in the photoelectric effect and pair production.

Answer is C.

NUCLEAR RADIATION–3

Which process requires a minimum of 1.02 MeV?

(A) II
(B) III
(C) VI
(D) VII

Solution:

Processes II and VI have been explained in Prob. 1

Process VII represents *isomeric transition* (IT). Most nuclei in excited states decay essentially immediately. If, due to the arrangement of their internal structure, an excited state's decay is delayed, the state appears as semistable or metastable. Such states are called *isomeric states*. When the decay occurs a photon is emitted from the nucleus and the entire process is called an isomeric transition.[7] Therefore, D is not the answer.

Pair production, process III, occurs in the vicinity of a strong coulomb field (i.e., near the nucleus). The photon disappears and an electron pair, a positron and an electron, appear.[8] The rest mass energy of two electrons is

$$E = 2mc^2$$

The rest mass of an electron is obtained from Table 3.1 in App. 3.A. The speed of light is 3.00 × 10[8] m/s.

---

[6] This is the equivalent of saying that the cross section σ_{cs} for Compton scattering is proportional to Z/E.
[7] The photon emitted from the nucleus is termed a gamma ray.
[8] The electron is called a negatron is some older texts.

Substituting gives

$$
\begin{aligned}
E &= 2mc^2 \\
&= (2)(9.109\times10^{-31}\ \text{kg})\left(3.00\times10^{8}\ \frac{\text{m}}{\text{s}}\right)^2 \\
&= \left(1.64\times10^{-13}\ \text{J}\right)\left(\frac{1\ \text{eV}}{1.602\times10^{-19}\ \text{J}}\right)\left(\frac{1\ \text{MeV}}{1\times10^{6}\ \text{eV}}\right) \\
&= 1.02\ \text{MeV}
\end{aligned}
$$

Therefore, the minimum required energy is 1.02 MeV. The probability for the occurrence of pair production is[9]

$$P_{\text{pp}} \propto Z^2$$

Z = atomic number

It should be understood that below 1.02 MeV the probability of pair production is zero.

Answer is B.

Nuclear Radiation (NR) problems 4 and 5 are based on the following information.

The total cross section for γ-ray interaction for the three most important effects with regard to shielding is

$$\sigma = \sigma_{\text{pe}} + \sigma_{\text{pp}} + \sigma_{\text{c}}$$

σ_{pe} = microscopic cross section for photoelectric effect
σ_{pp} = microscopic cross section for pair production
σ_{c} = microscopic cross section for Compton scattering[10]

Multiplying the cross section by the atom density N would normally give the macroscopic cross section Σ as the result.

By tradition, the macroscopic cross sections of gamma ray interaction are called *attenuation coefficients* and symbolized by μ.[11] Therefore,

$$\mu = N\sigma = \mu_{\text{pe}} + \mu_{\text{pp}} + \mu_{\text{c}}$$

When divided by the density of the material these coefficients become the *mass attenuation coefficients* and are routinely tabulated as such.[12]

$$\frac{\mu}{\rho} = \frac{\mu_{\text{pe}}}{\rho} + \frac{\mu_{\text{pp}}}{\rho} + \frac{\mu_{\text{c}}}{\rho}$$

To account for the energy deposited, which is closely related to the biological damage, the average energy of the recoiling electron in Compton scattering is accounted for and a *mass absorption coefficient* is determined.[13]

$$\frac{\mu_a}{\rho} = \frac{\mu_{\text{pe}}}{\rho} + \frac{\mu_{\text{pp}}}{\rho} + \frac{\mu_{\text{c},a}}{\rho}$$

The mass absorption coefficient is used to determine radiation intensity I and an energy deposition rate E_{dr}.

NUCLEAR RADIATION–4

Lead with a mass thickness of 86.49 g/cm^2 is to be used to shield against 1.33 MeV γ-rays of intensity $3\times10^5\ \text{s}^{-1}$.[14]

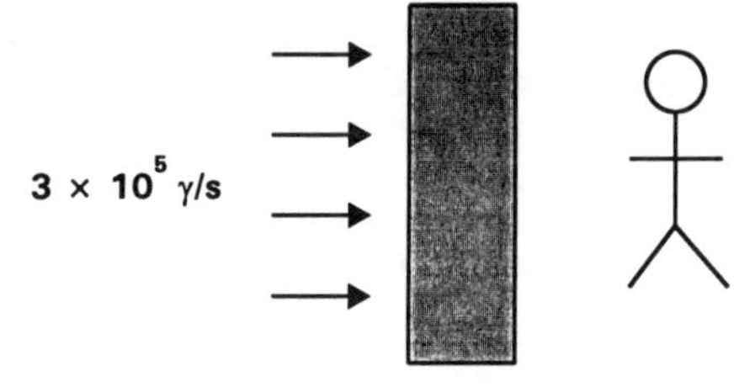

[9] Again, this is equivalent to saying that the cross section σ_{pp} is proportional to the atomic number Z squared.

[10] In the Compton scattering interaction the photon (or γ-ray) interacts with individual electrons. Therefore, a cross section per electron σ_{ce} exists. The total probability for any given atom is the number of electrons in the atom Z_e multiplied by the individual probabilities σ_{ce}.. In other words, $\sigma_c = Z_e\sigma_{\text{ce}}$.

[11] This represents the probability per unit path length of an interaction (i.e., photoelectric effect (PEE); pair production (PP) and Compton scattering (CS)). Physicists call these items linear absorption coefficients. Not all the gamma rays are absorbed however, and this must be taken into account in shielding problems by using buildup factors.

[12] These are useful in that the ratio μ/ρ is essentially constant over the range of energies where Compton scattering dominates (i.e., from approximately 0.5 MeV to 5 MeV) and thus the effectiveness of a shield varies only with its mass thickness (i.e., ρx). See the Reference section for more information.

[13] These are called energy absorption coefficients in some books and given the symbol μ_{en}.

[14] The γ-ray from cobalt 60 decay has an energy of approximately 1.33 MeV.

What is the approximate intensity of the gamma rays near the worker which have not interacted with the lead?[15]

(A) 2.6×10^2 s^{-1}
(B) 2.2×10^3 s^{-1}
(C) 3.0×10^5 s^{-1}
(D) 4.0×10^7 s^{-1}

Solution:

First, it should be noted the position of the worker behind the lead in this case is of little consequence. The reasons are that the non-interacting photons are not expanding as from a point source; therefore, distance will not lower the intensity of the beam as would normally occur from simple spherical-like expansion of the γ-rays from their point of origin. Importantly, the mass attenuation coefficient for γ-rays in air is insignificant for most forms of interaction and can be ignored for many tens of meters depending upon the accuracy desired.[16]

The applicable formula is

$$I = I_0 e^{-\left(\frac{\mu}{\rho}\right)(\rho x)}$$

I_0 = initial intensity
μ/ρ = mass attenuation coefficient
ρx = mass thickness; the density of the shield material multiplied by the thickness[17]

The mass attenuation coefficient for lead interacting with a 1.33 MeV gamma is obtained from Table 3.2 of App. 3.A. Substituting gives

$$I = I_0 e^{-\left(\frac{\mu}{\rho}\right)(\rho x)}$$

$$= \left(3 \times 10^5 \frac{\gamma}{\text{s}}\right) \exp\left(\left(-5.67 \times 10^{-2} \frac{\text{cm}^2}{\text{g}}\right)\left(86.49 \frac{\text{g}}{\text{cm}^2}\right)\right)$$

$$= \boxed{2.2 \times 10^3 \ \gamma/\text{s}}$$

Answer is B.

NUCLEAR RADIATION–5

A 15 kg carbon steel drum comprised of 95 w/o (i.e., weight percent) iron and 5 w/o silicon is used to store a radioactive liquid.[18] The activity of the liquid on the interior surface of the drum is estimated to be approximately 2×10^{11} γ/cm^2·s with an average gamma ray energy of 1.0 MeV.

What is the energy deposition rate into the drum?

(A) 1–5 W
(B) 5–10 W
(C) 10–15 W
(D) 15–20 W

Solution:

The energy deposition rate per unit mass is

$$E_{\text{DR}} = E_\gamma I \left(\frac{\mu_a}{\rho}\right)$$

E_γ = photon energy
I = intensity
μ_a/ρ = mass absorption coefficient

Since the steel is comprised of two materials, the overall mass absorption coefficient can be calculated from

$$\left(\frac{\mu_a}{\rho}\right)_{\text{steel}} = \left(w_{\text{Fe}}\left(\frac{\mu_a}{\rho}\right)_{\text{Fe}} + w_{\text{Si}}\left(\frac{\mu_a}{\rho}\right)_{\text{Si}}\right)(0.01)$$

w_x = weight percent of the given material
0.01 = conversion of percent to decimal

[15] The gamma rays that have not interacted are not the only photons present on the worker's side. Compton scattering results in scattered photons, *x*-rays may follow the photoelectric effect, and annihilation radiation follows pair production, all which may be present depending upon where they were generated in the lead. These effects can be factored into the appropriate equations using buildup factors as will be seen in follow-on problems. This question is a rough calculation since photons that have interacted have been ignored; thus, the need to use a buildup factor has been eliminated.

[16] The value of the mass attenuation coefficient for air is approximately 10^{-4} cm^2/g for 1.3 MeV gammas.

[17] This combination ρx results in units of mass/area (e.g., g/cm^2) which is a common way of expressing shield thickness.

[18] Silicon makes the steel harder. Its main purpose here, however, is as the impetus to explain the use of the mass absorption coefficient in mixtures of material.

The values for the mass absorption coefficients are found in Table 3.3 of Appendix 3.A. Substituting gives

$$\begin{aligned}\left(\frac{\mu_a}{\rho}\right)_{\text{steel}} &= \left(w_{\text{Fe}}\left(\frac{\mu_a}{\rho}\right)_{\text{Fe}} + w_{\text{Si}}\left(\frac{\mu_a}{\rho}\right)_{\text{Si}}\right)(0.01)\\ &= (95)\left(2.603\times10^{-2}\ \frac{\text{cm}^2}{\text{g}}\right)_{\text{Fe}}(0.01)\\ &\quad+(5)\left(2.778\times10^{-2}\ \frac{\text{cm}^2}{\text{g}}\right)_{\text{Si}}(0.01)\\ &= 2.612\times10^{-2}\ \text{cm}^2/\text{g}\end{aligned}$$

Calculate the energy deposition rate per unit mass.

$$\begin{aligned}E_{\text{DR}} &= E_\gamma I\left(\frac{\mu_a}{\rho}\right)_{\text{Steel}}\\ &= (1.0\ \text{MeV})\left(2\times10^{11}\ \frac{\gamma}{\text{cm}^2\cdot\text{s}}\right)\left(2.612\times10^{-2}\ \frac{\text{cm}^2}{\text{g}}\right)\\ &= 5.22\times10^{9}\ \text{MeV/g}\cdot\text{s}\end{aligned}$$

Determine the rate for the entire drum in the requested units.

$$\begin{aligned}E_{\text{DR,total}} &= E_{\text{DR}}m\\ &= \left(5.22\times10^{9}\ \frac{\text{MeV}}{\text{g}\cdot\text{s}}\right)(15\ \text{kg})\left(\frac{1\times10^{3}\ \text{g}}{\text{kg}}\right)\\ &= \left(7.84\times10^{13}\ \frac{\text{MeV}}{\text{s}}\right)\left(\frac{1\times10^{6}\ \text{eV}}{\text{MeV}}\right)\left(1.602\times10^{-19}\ \frac{\text{J}}{\text{eV}}\right)\\ &= 12.6\ \frac{\text{J}}{\text{s}}\\ &= \boxed{12.6\ \text{W}}\end{aligned}$$

Answer is C.

Nuclear Radiation (NR) problems 6 and 7 are based on the following information.

selected properties of cobalt	
density Co-59	8.8 g/cm³
σ_a	37.2 barns
AW	58.9332
Co-60 half-life	5.271 yr

NUCLEAR RADIATION–6

A small cylindrical support pin with a radius of 0.5 cm and length of 1.0 cm is made from Co-60. The pin is exposed to a monoenergetic neutron flux.

$$\phi = 8\times10^{5}\ \text{neutrons/cm}^2\cdot\text{s}$$

In one year of steady state operation, what is the number of atoms activated in this pin?

(A) 2.7×10^{6} atoms
(B) 6.6×10^{13} atoms
(C) 9.0×10^{13} atoms
(D) 12.9×10^{20} atoms

Solution:

Cobalt 59 is activated when it absorbs a neutron and becomes Co-60.

$${}^{59}_{27}\text{Co}+{}^{1}_{0}\text{n}\rightarrow{}^{60}_{27}\text{Co}$$

Therefore, calculate the number of collisions that occur where a neutron is absorbed. The collision density can be calculated for the desired type of reaction (i.e., absorption, scattering, capture and so on) from the general formula

$$F = IN\sigma = I\Sigma$$

F = collision density [collisions/volume·time]
I = intensity [neutrons/area·time][19,20]
N = atom density
σ = microscopic cross section for the desired event

[19] The formula for the intensity is $I = nv$ and has units of radiation particles/volume·velocity. If the intensity I is equal to a flux ϕ then the formula becomes $F = \Sigma\phi$ as long as the neutrons are monoenergetic. See the Nuclear Theory section for more information.

[20] This could also be for gamma rays, beta rays, and so on. It represents the radiation beam or flux intensity.

For neutron absorption this formula becomes

$$F = IN_{\text{Co-59}}\sigma_a$$

The atom density of the Co-59 is the only unknown. It can be calculated from

$$\begin{aligned}N_{\text{Co-59}} &= \frac{\rho N_A}{\text{AW}}\\ &= \frac{\left(8.8\ \frac{\text{g}}{\text{cm}^3}\right)\left(6.02\times 10^{23}\ \frac{\text{atoms}}{\text{mol}}\right)}{58.9332\ \frac{\text{g}}{\text{mol}}}\\ &= 8.99\times 10^{22}\ \text{atoms/cm}^3\end{aligned}$$

Substituting this atom density into the collision formula gives

$$\begin{aligned}F &= IN\sigma_a\\ &= \left(8\times 10^5\ \frac{\text{neutrons}}{\text{cm}^2\cdot\text{s}}\right)\left(8.99\times 10^{22}\ \frac{\text{atoms}}{\text{cm}^3}\right)\\ &\quad\times(37.2\ \text{barns})\left(\frac{1\times 10^{-24}\ \text{cm}^2}{1\ \text{barn}}\right)\\ &= 2.68\times 10^6\ \text{collisions/cm}^3\cdot\text{s}\end{aligned}$$

This is the number of collisions that took place in this pin (which is why the atom density of Co-59 was used) that resulted in the absorption of a neutron (which is why the microscopic cross section for absorption σ_a was used) and therefore represents the activation rate.

For the entire year,

$$\begin{aligned}F_{\text{total}} &= Ft\\ &= \left(2.68\times 10^6\ \frac{\text{collisions}}{\text{cm}^3\cdot\text{s}}\right)(1\ \text{yr})\left(\frac{365\ \text{days}}{1\ \text{yr}}\right)\\ &\quad\times\left(\frac{24\ \text{hr}}{1\ \text{day}}\right)\left(\frac{3600\ \text{s}}{1\ \text{hr}}\right)\\ &= 8.44\times 10^{13}\ \text{atoms/cm}^3\end{aligned}$$

The pin is a cylinder; therefore, the volume is

$$\begin{aligned}V &= \pi r^2 l\\ &= \pi(0.5\ \text{cm})^2(1\ \text{cm})\\ &= 0.785\ \text{cm}^3\end{aligned}$$

The total number of atoms activated in the year is

$$\begin{aligned}\text{atoms activated} &= F_{\text{total}}V\\ &= \left(8.44\times 10^{13}\ \frac{\text{atoms}}{\text{cm}^3}\right)(0.785\ \text{cm}^3)\\ &= \boxed{6.63\times 10^{13}\ \text{atoms}\quad (6.6\times 10^{13}\ \text{atoms})}\end{aligned}$$

Answer is B.

NUCLEAR RADIATION–7

What is the approximate activity of the pin in the Prob. 6 at the one year point, in becquerels?

(A) 2.76×10^5 Bq
(B) 3.55×10^5 Bq
(C) 8.68×10^{12} Bq
(D) 1.18×10^{13} Bq

Solution:

The decay rate is referred to as the activity. In general form,

$$\alpha(t) = \lambda\, n(t)$$

$\alpha(t)$ = activity in Bq (disintegrations/s)
λ = decay constant
$n(t)$ = undecayed atoms

The activity at the one year point was requested and the half-life $T_{1/2}$ is given; therefore,[21]

$$\begin{aligned}\alpha &= \left(\frac{0.693}{T_{1/2}}\right)n\\ &= \left(\frac{0.693}{5.271\ \text{yr}}\right)(6.63\times 10^{13}\ \text{atoms})\\ &= 8.72\times 10^{12}\ \text{disintegrations/yr}\end{aligned}$$

[21] If this were not the case, an equilibrium equation would need to be used, balancing the production and decay. See the Reference section for texts which explain this type of process.

Change the units to disintegrations/s and note that a becqueral is one disintegration/s. This gives the following result.[22]

$$\alpha = 8.72 \times 10^{12}\ \frac{\text{disintegrations}}{\text{yr}}$$

$$= \left(8.72 \times 10^{12}\ \frac{\text{disintegrations}}{\text{yr}}\right)\left(\frac{1\ \text{yr}}{3.16 \times 10^{7}\ \text{s}}\right)$$

$$\times \left(\frac{1\ \text{Bq}}{1\ \dfrac{\text{disintegration}}{\text{s}}}\right)$$

$$= \boxed{2.76 \times 10^{5}\ \text{Bq}}$$

Answer is A.

NUCLEAR RADIATION–8

A sample of primary coolant is drawn from a pressurized water reactor (PWR). It is depressurized and the gaseous constituents contained in an enclosed chamber with a 2 mm steel cover on one side which allows for activity measurements. The initial activity measurement A_0 is 111 counts per second (cps) above normal.

In discussions the following potential causes are surmised.

(A) The activity is a result of manganese and iron from corrosion products due to the shock from a recent startup following an extended refueling.

(B) The activity is due to naturally occurring nitrogen from the following reaction.[23]

$$^{16}\text{O}[\text{n, p}]^{16}\text{N}$$

(C) The activity may be from xenon 135 indicating a potential fuel element failure may have occurred while shutdown.

(D) During the recent pressurization following refueling, air may have been added to the plant resulting in argon 41 activity.

Following the discussion, four hours after the initial sample, a second activity is taken on the original sample and a γ-ray spectrometer measurement is taken.

$$A_4 = 22\ \text{cps}$$

$$\text{detected gamma} = 1.3\ \text{MeV}\ \gamma$$

What is the source of the activity?

(A) Mn-56 and Fe-59
(B) N-16
(C) Xe-135
(D) Ar-41

Solution:

The sample is gaseous in nature. Both Mn-56 and Fe-59 are solids or corrosion particulates in a liquid. Thus, A is not the answer.

The half-life of N-16 is 7.13 seconds as seen in Table 3.4 of App. 3.A. It is the primary contributor to radiation levels during operation of a reactor plant. Further, its high energy gamma rays are biologically significant and of major concern in shielding calculations. Nevertheless, in five half-lives most of the activity (i.e., over 97%) would be gone; therefore, it would not account for the activity level at the four hour point. Thus, B is not the answer.

The Xe-135 is a gas, and if present in excessive quantities, does indeed indicate a potential fuel element failure. Nevertheless, from Table 3.4 of App. 3.A, its half-life is 9.10 hr and the gamma emitted has an energy of 0.3 MeV. Therefore, these properties don't match the measured values. Thus, C is not the answer.[24]

---

[22] The value shown for seconds in a year differs slightly from that previously used. Be aware that advanced calculators use 365.25 days in a year in conversion programs; this is the reason for the slight difference.

[23] The symbology shown indicates an oxygen 16 atom absorbing a neutron and then decaying by expulsion of a proton to become nitrogen 16.

[24] The β^- decays of all of these nuclides have not been mentioned. The range for beta particles of the approximate energy emitted by these nuclides is on the order of 1 mm in steel. Thus, the 2 mm cover on the sample container would block these particles.

A graph of the activity measurements follows.

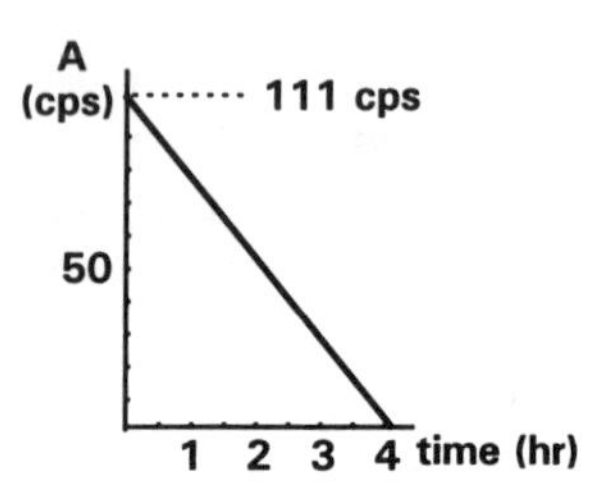

A straight line interpolation indicates a half-life of approximately two hours. A more accurate measurement using an exponential decay curve shows a half-life of 1.8 hours.

Argon-41, from Table 3.4 of App. 3.A has a half-life of 1.82 hr and emits a gamma of 1.3 MeV. Further, Ar-40 comprises about 3% of air and it's possible that during pressurization air was charged into the plant.[25] The Ar-40 is activated by neutron absorption to Ar-41, from which it decays giving the properties noted.

Answer is D.

Nuclear Radiation (NR) problems 9–12 are based on the following information which represent a portion of a chart of the nuclides.[26]

Z ↑	N →	
Ba-136	Ba-137	Ba-138
Cs-135	Cs-136	Cs-137
Xe-134	Xe-135	Xe-136

N = number of neutrons
Z = number of protons

Now consider the following portion of the chart of the nuclides. The shaded block represent the original nucleus. The other blocks are arbitrarily labeled to represent the indicated position on the chart.

			IX	XIII
	III	VII	X	XIV
	IV	(shaded)	XI	
I	V	VIII	XII	
II	VI			

NUCLEAR RADIATION–9

If the original nucleus decays by the emission of an alpha particle, what position will it then occupy on the chart?

(A) II
(B) V
(C) X
(D) XIII

Solution:

An alpha particle consists of two neutrons and two protons and is often represented by the following symbol.[27]

$$_2\alpha^4 \text{ or } {}^4_2\text{He}$$

Such an emission would result in a movement two positions to the left and two positions down from the original (i.e., to position II).

Position V represents the expulsion of a deuteron (designated by a "d" in the charts) and sometimes shown as the following.[28]

$${}^2_1\text{H}$$

Position X represents the absorption of a deuteron. Position XIII represents the absorption of an alpha particle.

Answer is A.

[25] It should be noted that several tens of liters of air would be required to generate such activity for an average size commercial electrical generation plant.

[26] An actual chart contains significantly more information. It is recommended one become familiar with such a chart prior to the nuclear profession engineer examination. See the Reference section for more information.

[27] One should remember that if the helium designation is used no electrons are attached to an alpha particle (or helium nucleus) when it is emitted.

[28] Again, no electrons are attached upon emission. Further, when an electron becomes attached, it is referred to as a deuterium atom. The nucleus alone is called a deuteron.

NUCLEAR RADIATION–10

If the original nucleus decays by β^- emission, what position will it then occupy on the chart?

(A) III
(B) IV
(C) VII
(D) XI

Solution:

Normally a beta decay occurs to increase the stability of a nucleus whose neutron to proton ratio is not optimum.[29] A neutron emits a beta particle (essentially an electron) and becomes a proton (with charge conservation at work). Therefore, the original nucleus loses a neutron (i.e., moves one position to the left) and gains a proton (i.e., moves one position up). The final position is III.

Position IV represents a neutron loss, position XI a neutron gain. Position VII represents a proton expulsion.

Answer is A.

NUCLEAR RADIATION–11

Consider the following portion of a chart of the nuclides.

Z ↑				
	U-232	U-233	U-234	U-235
	Pa-231	Pa-232	Pa-233	Pa-234
	Th-230	Th-231	Th-232	Th-233
Z	Ac-229	Ac-230	Ac-231	Ac-232

N ⟶

The original nucleus is U-235. The U-235 undergoes alpha decay; the resulting nucleus β^- decays. This nucleus then absorbs two neutrons. What is the resultant element?

(A) Ac-231
(B) Th-232
(C) Pa-233
(D) U-234

[29] Neutrons and their associated nuclear forces balance the protons' repulsive electromagnetic forces. Maximum stability occurs with a spherical charge distribution which can be detected by measuring zero nuclear electric quadrupole moments. Those nuclei with what is termed a magic number of neutrons and protons (e.g., 2, 8, 20, 28, 50, 82, and 126) are the most stable. See a modern physics text for additional information. See the Reference section for an example.

Solution:

First, the elements involved are actinium, thorium, protactinium and uranium. The chain of events given is a probable one based on the data in a complete chart of the nuclides.

Uranium-235 is the starting point. Following alpha decay, Th-231 exists. The subsequent β^- decay results in Pa-231 which has a very high cross section for neutron absorption. Absorbing one neutron gives Pa-232 and a second gives Pa-233.

Answer is C.

NUCLEAR RADIATION–12

Consider the symbology introduced earlier.

$$X[\text{particle absorbed; particle emitted}]Y$$

X = original nucleus
Y = resulting nucleus

If the original nucleus undergoes a neutron capture reaction [n,γ], what is its final position on the chart?

(A) I
(B) IV
(C) XI
(D) XII

Solution:

In this case nuclear reactions are considered instead of decays. The principle, regarding use of the chart, is the same.

Position I represents a [p,α] reaction. Position IV can represent [p,pn], [γ,n], or [n,2n]. Position XII includes the reactions [n,p] or [t,He-3]. Position XI represents [d,p], [n,γ], or [t,np] and thus contains the desired result.[30] Specifically, the absorption moves the nucleus one position to the right; the gamma lowers the energy but has no effect on the number of protons or neutrons.

Answer is C.

[30] Recall that d is the symbology for a deuteron. The t represents a triton (i.e., a hydrogen-3 nucleus).

NUCLEAR RADIATION–13

Consider the following graph and table.[31]

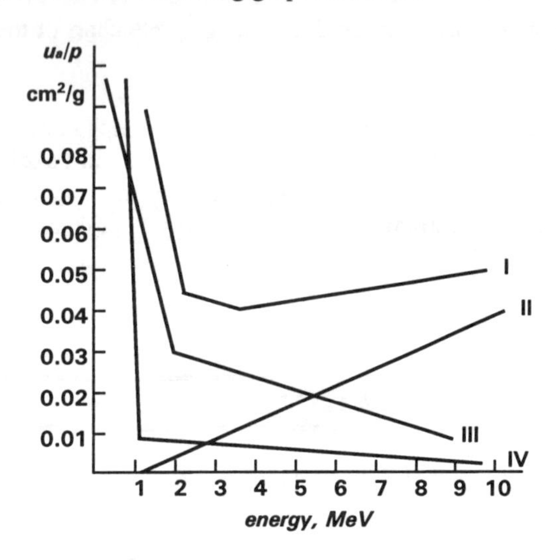

Match each line (i.e., I, II, III, IV) with the associated effect given below.

designation	effect
A	photoelectric effect
B	Compton scattering
C	pair production
D	total

(A) I A II B III C IV D
(B) I A II C III B IV D
(C) I D II B III A IV C
(D) I D II C III B IV A

Solution:

Curve I represents the total as can be seen from mental addition of the lines below it. Thus, the correct designation is I-D.

Curve II doesn't begin until after an energy of approximately 1.0 MeV and increases as the energy becomes larger. This represent pair production. Thus the correct designation is II-C.

Curve III represents an effect which dominates in the mid-range of gamma ray energy and thus is the mass absorption curve for Compton scattering. The correct designation is III-B.

Curve IV represents the dominate effect for low energy gamma rays and thus is the mass absorption curve for the photoelectric effect. The designation is IV-A.

The final result is I-D, II-C, III-B, and IV-A.

Answer is D.

Nuclear Radiation (NR) problems 14 and 15 are based on the following information.

traditional unit	symbol	unit of	SI unit	notes
roentgen[32]	R[a]	exposure	C/kg	1 R = 2.58×10^{-4} C/kg
rad	rad	absorbed dose	Gy	1 gray = 1 J/kg = 100 rad
rem	rem	absorbed dose equivalent	Sv	1 sievert = 100 rem

[a] The symbol R used for roentgen is often also used to indicate rem. Caution is advised. In the SI system, this problem does not exist.

quality factors	
x-ray; γ-ray; β-ray	1
thermal neutrons	2
neutrons with unknown energy; protons	10
alpha particles	20

NUCLEAR RADIATION–14

A 10 mR/hr 1.0 MeV gamma radiation field exists in a localized area of a nuclear power plant. At approximately what rate are ions being produced in the air in this area?

(A) 3×10^{-7} cm$^{-3}\cdot$s^{-1}
(B) 30×10^{-7} cm$^{-3}\cdot$s^{-1}
(C) 60 cm$^{-3}\cdot$s^{-1}
(D) 6000 cm$^{-3}\cdot$s^{-1}

[31] The graph is shown with linear approximations for ease of use only. See the Reference section for texts with more detailed information.

[32] Additional useful information follows. One *esu* = 3.33×10^{-10} C. The mass of one cm³ of dry air at STP = 0.001293 g.

Solution:

At an energy of 1 MeV the photoelectric effect dominates with some contribution from Compton scattering. Assuming any scattered gamma from Compton scattering eventually goes on to cause the photoelectric effect, only singly charge ions are produced.[33] Each ion has a charge of 1.60×10^{-19} C.

Determine the charge associated with a roentgen. The definition of a roentgen, prior to the adoption of the SI system, was that amount of radiation which will produce one electrostatic unit (i.e., esu) of charge in one cubic centimeter (cm^3) of dry air at standard temperature and pressure (STP; 1 atmosphere and 0°C).

Since one esu equal 3.33×10^{-10} C and 1 cm^3 of air at STP is 0.001293 g,

$$\begin{aligned} 1R &= \frac{1\,\text{esu}}{m_{\text{air,cm}^3}} \\ &= \frac{3.33\times 10^{-10}\,\text{C}}{0.001293\,\text{g}} \\ &= \left(2.58\times 10^{-7}\,\frac{\text{C}}{\text{g}}\right)\left(\frac{1\times 10^{3}\,\text{g}}{1\,\text{kg}}\right) \\ &= 2.58\times 10^{-4}\ \text{C/kg} \end{aligned}$$

Note: This can also be taken directly from the given table but was calculated here for clarification.

Combining this information to determine the ion production rate per unit volume gives

$$\begin{aligned} \dot{I} &= (\text{exposure rate})\left(\frac{\text{charge / unit mass of air}}{\text{unit of exposure}}\right) \\ &\quad\times\left(\frac{\text{ions}}{\text{charge}}\right)\left(\frac{\text{mass}}{\text{volume}}\right) \\ &= \left(10\times 10^{-3}\,\frac{\text{R}}{\text{hr}}\right)\left(\frac{2.58\times 10^{-4}\,\frac{\text{C}}{\text{kg}}}{1\,\text{R}}\right)\left(\frac{1\,\text{ion}}{1.602\times 10^{-19}\,\text{C}}\right) \\ &\quad\times\left(\frac{0.001293\,\text{g}}{1\,\text{cm}^3}\right)\left(\frac{1\,\text{kg}}{1\times 10^{3}\,\text{g}}\right)\left(\frac{1\,\text{hr}}{3600\,\text{s}}\right) \\ &= \boxed{5.79\times 10^{3}\ \text{ions/cm}^3\cdot\text{s}\quad (6000\,\text{cm}^{-3}\cdot\text{s}^{-1})} \end{aligned}$$

Answer is D .

NUCLEAR RADIATION–15

What is the dose equivalent for a person who remains in the gamma ray field described in the Prob. 14 for 8 hours?

(A) 80 mGy
(B) 800 μGy
(C) 80 mSv
(D) 800 μSv

Solution:

The dose equivalent and dose equivalent rate are defined as follows.

$$\text{dose equivalent} = (\text{absorbed dose})(\text{quality factor})$$
$$H = DQ$$
$$\text{dose equivalent rate} = (\text{absorbed dose rate})(\text{quality factor})$$
$$\dot{H} = \dot{D}Q$$

D = energy deposition per unit mass
Q = weighting factor which accounts for the biological effects of different types of radiation[34]

[33] Even without this assumption, at 1.0 MeV the mass attenuation coefficient for the photoelectric effect is roughly 1000 times larger that the same coefficient for Compton scattering.

[34] The quality factor is related to the linear energy transfer (LET). See the Reference section for additional information.

The gray Gy is a unit of absorbed dose, not the dose equivalent, so A and B are not the answer.[35]

Since an exposure to one roentgen of gamma rays at 1.0 MeV results in a deposition of 0.96×10^{-2} J/kg of soft body tissue, one rad of absorbed dose occurs from approximately one roentgen of exposure.[36,37] Thus, the absorbed dose rate is 10 mrad/hr. Additionally, the quality factor for gamma rays is one, as taken from the information table provided prior to this problem. Substituting,

$$\begin{aligned}\dot{H} &= \dot{D}Q \\ &= \left(10\,\frac{\text{mrad}}{\text{hr}}\right)(1) \\ &= 10\ \text{mrem/hr}\end{aligned}$$

The dose equivalent is

$$\begin{aligned}H &= \dot{H}t \\ &= \left(10\,\frac{\text{mrem}}{\text{hr}}\right)(8\ \text{hr}) \\ &= (80\ \text{mrem})\left(\frac{1\ \text{rem}}{1\times 10^{3}\ \text{mrem}}\right)\left(\frac{1\ \text{Sv}}{100\ \text{rem}}\right) \\ &= 8.00\times 10^{-4}\ \text{Sv} \\ &= \boxed{800\ \mu\text{Sv}}\end{aligned}$$

Answer is D.

[35] The dose equivalent accounts for the biological effects.

[36] This is an important concept. This assumption allows the exposure taken directly from instruments, calibrated to measure in roentgens, to be directly related on a one-to-one basis with the absorbed dose in biological tissue. Of note, this is not the true for bone, as the absorbed dose is higher for a one roentgen exposure.

[37] Knowing the energy deposition rate, which is determined by experiment, one can calculate the exposure rate from

$$\dot{X} = \frac{C I_\gamma E_\gamma \left(\dfrac{\mu_a}{\rho}\right)_{\text{air}}}{0.96\times 10^{-2}\ \dfrac{\text{J}}{\text{kg}}}$$

The constant C depends on the desired units. See the texts in the Reference section for additional information.

Nuclear Radiation (NR) problems 16–18 are based on the following information.

Consider the following potential shielding materials.

designation	material
I	iron
II	lead
III	water
IV	paraffin
V	polyethylene

Consider the following list of shield material advantages.

designation	advantage
a	inexpensive
b	high density
c	high strength
d	no neutron activation problems

Consider the following list of shield material disadvantages.

designation	disadvantage
i	activates in neutron fluxes
ii	toxic
iii	combustible; fuel for a fire
iv	potential loss of shield due to leakage

NUCLEAR RADIATION–16

What are the advantages of lead (II) and polyethylene (V) for use as shielding?

(A) II a V a
(B) II b V d
(C) II c V a
(D) II d V d

Solution:

The following table groups the material with its advantages.

designation	material	advantages
I	iron	a,c
II	lead	a,b
III	water	a
IV	paraffin	a,d
V	polyethylene	d

Therefore II-b and V-d is the correct combination.

Answer is B.

NUCLEAR RADIATION–17

What are the disadvantages of iron (I) and paraffin (IV)?

(A) I i IV iii
(B) I ii IV iii
(C) I iii IV iv
(D) I iv IV iv

Solution:

The following table groups the material with its disadvantages.

designation	material	disadvantages
I	iron	i
II	lead	ii
III	water	iv
IV	paraffin	iii
V	polyethylene	iii

Therefore I-i and IV-iii is the correct combination.

Answer is A.

NUCLEAR RADIATION–18

What is a tenth-value layer for the uncollided flux in ordinary concrete for a 1.0 MeV gamma ray beam?[38] The density of ordinary concrete is approximately 2.3 g/cm^3.

(A) 5 cm
(B) 10 cm
(C) 15 cm
(D) 20 cm

Solution:

The applicable formula, the terms of which were given in earlier problems, is[39]

$$I = I_0 e^{-\left(\frac{\mu}{\rho}\right)(\rho X)}$$

Rearranging to determine the tenth-value gives

$$I = I_0 e^{-\left(\frac{\mu}{\rho}\right)(\rho X)}$$

$$0.1 = \exp\left(-\left(\frac{\mu}{\rho}\right)(\rho X)\right)$$

$$\ln(0.1) = -\left(\frac{\mu}{\rho}\right)\rho X$$

$$X = \frac{\ln(0.1)}{-\left(\frac{\mu}{\rho}\right)\rho}$$

The mass attenuation coefficient μ/ρ, from Table 3.2 in App. 3.A is

$$\frac{\mu}{\rho} = 6.495 \times 10^{-2}\ \text{cm}^2/\text{g}$$

The density is given as 2.3 g/cm^3.

[38] The tenth-value layer is also called the tenth-thickness. Half-value thicknesses are commonly tabulated as well. The reduction of a given beam of radiation to five percent of its original value is also a common reference.

[39] The use of the uncollided flux indicates that the buildup factor can be ignored. The result, therefore, is approximate and underestimates the actual thickness required.

Substituting gives

$$X = \frac{\ln(0.1)}{-\left(\frac{\mu}{\rho}\right)\rho}$$

$$= \frac{\ln(0.1)}{-\left(6.495\times10^{-2}\ \frac{\text{cm}^2}{\text{g}}\right)\left(2.3\ \frac{\text{g}}{\text{cm}^3}\right)}$$

$$= \boxed{15.4\ \text{cm} \quad (15\ \text{cm})}$$

Answer is C .[40]

NUCLEAR RADIATION–19

An isotropic point source is used as a calibration test source for radiation detection instruments. It is stored in a lead container as shown.

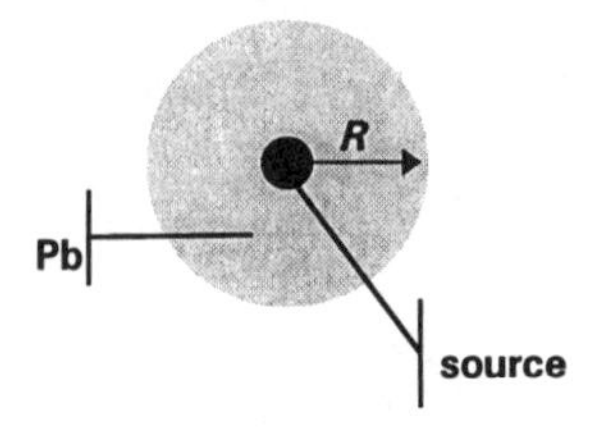

$R = 1$ cm
$\rho_{Rb} = 11.35$ g/cm^3
$E_\gamma = 1.0$ MeV

A quick calculation assuming spherical spreading of the gamma flux indicates the exposure rate to be 10 mR/hr without the shield present.[41]

Using the Taylor form of the point isotropic buildup factor, what is the expected exposure rate at the surface of thc lead shield?

(A) 2 mR/hr
(B) 5 mR/hr
(C) 6 mR/hr
(D) 10 mR/hr

Solution:

The applicable formula for the exposure rate is

$$\dot{X} = \dot{X}_0 B_p(\mu R)e^{-\mu R}$$

The Taylor form of the buildup factor is as follows.

$$B_p(\mu R) = A_1 e^{-\alpha_1 \mu R} + A_2 e^{-\alpha_2 \mu R}$$

The values for A_1, A_2, α_1, and α_2 are taken from Table 3.5 of App. 3.B. The radius R is given as 1 cm. The value for the mass attenuation coefficient for 1.0 MeV gammas is taken from Table 3.2 of App. 3.A. Thus,

$$\mu = \left(\frac{\mu}{\rho}\right)_{PB} \rho_{PB}$$

$$= \left(6.84\times10^{-2}\ \frac{\text{cm}^2}{g}\right)\left(11.35\ \frac{\text{g}}{\text{cm}^3}\right)$$

$$= 0.776\ \text{cm}^{-1}$$

Substitute to determine the point isotropic buildup factor.

$$B_p(\mu R) = A_1 e^{-\alpha_1 \mu R} + A_2 e^{-\alpha_2 \mu R}$$

$$= (2.984)\exp\left(-(-0.03503)(0.776\ \text{cm}^{-1})(1\ \text{cm})\right)$$

$$+ (1-2.984)\exp\left(-(0.13486)(0.776\ \text{cm}^{-1})(1\ \text{cm})\right)$$

$$= 1.28$$

Using this and the given information, calculate the exposure rate.

[40] The actual value required is 20 cm. Therefore, the use of buildup factors is imperative for accurate calculations on actual shields.

[41] An operational way of determining this exposure would be to simply measure it at the desired distance. One would then calculate the thickness of the shielding required for a desired exposure rate. A programmable calculator or semi-log paper is helpful. See the Reference section for texts which provide further examples of calculations of this sort.

$$\dot{X} = \dot{X}_0 B_p(\mu R)e^{-\mu R}$$
$$= \left(10\frac{\text{mR}}{\text{hr}}\right)(1.28)\exp(-0.776)$$
$$= \boxed{5.89\ \text{mR/hr}\quad(6\ \text{mR/hr})}$$

Answer is C.[42]

NUCLEAR RADIATION–20

A radioactive spill occurs on the shielded floor of a sampling laboratory. The situation is as shown.

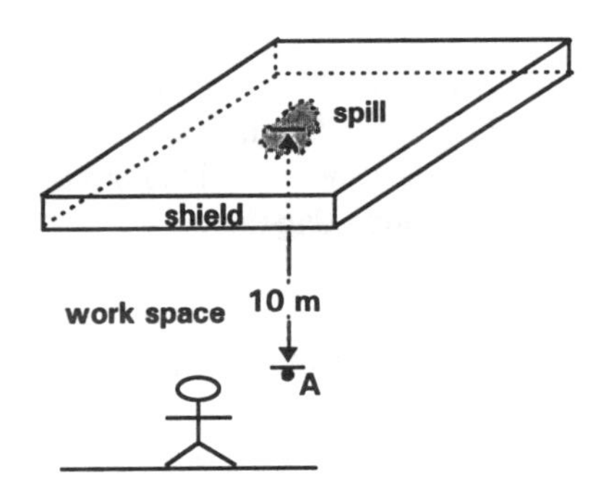

Assuming the spill is approximately circular, the solution for the absorbed dose equivalent at point A is of which form?

(A) 1/(10 m)
(B) Sievert integral[43]
(C) $1/(10\text{ m})^2$
(D) E_1 [exponential integral]

Solution:

Unshielded line source fluxes and exposures tend to decrease as 1/distance. Therefore, A is not the answer.

Shielded line sources fluxes and exposures take on the form of a Sievert integral function.

$$F(\theta, x) = \int_0^\theta e^{-x\sec\theta}\,d\theta$$

Therefore, B is not the answer.

Point source fluxes and exposures tend to decrease as a function of $1/\text{distance}^2$. Unshielded disc (i.e., planer) and ring source fluxes and exposures also tend to decrease as functions of $1/\text{distance}^2$. Therefore, C is not the answer.

Shielded disc or planar sources, which this problem represents, are of the form of an exponential integral with n = 1.

$$E_n(x) = x^{n-1}\int_x^\infty \frac{e^{-t}}{t^n}\,dt$$

The air between the shield and point A offers no significant attenuation for gammas and is usually ignored.[44]

Answer is D.[45]

Note: Other methods not discussed for shield design include removal-attenuation calculation; removal-diffusion (i.e., removal-attenuation with multiple neutron energy groups); transport and Monte Carlo. The latter two methods are considered exact methods and are used for both gamma and neutron radiation.

[42] The method used in this problem can be reversed to determine the shielding thickness for a desired exposure rate. First determine the *buildup flux* for the desired exposure rate. Using the applicable formula for the flux as determined by the point kernel technique, determine the number of mean free paths required (i.e., μR). Then solve for the thickness R required. See the Reference section for additional information. See the Nomenclature section for a definition of the above terms.

[43] This is also called the secant integral.

[44] The air is of little consequence up to 30 m or more depending upon the accuracy desired.

[45] Numerous concepts have been introduced here which are worthy of further study by a nuclear engineer.

NUCLEAR RADIATION–21

One millicurie of naturally occurring Ra-226 finds its way into a 5 000 000 gallon community water supply tank. Does the resulting water violate EPA standards for drinking water and what is the primary hazard from this element?[46]

(A) yes; internal α
(B) yes; external γ
(C) no; internal α
(D) no; external γ

Solution:

First, from Table 3.6 in App. 3.B, the concentration limit for Ra-226 is 0.2 Bq/L. Calculating the activity concentration of radium in the water supply gives the following.

$$[\mathrm{Ra}] = \frac{\text{activity}}{\text{volume}}$$

$$= \left(\frac{1\times10^{-3}\ \mathrm{Ci}}{5\,000\,000\ \mathrm{gal}}\right)\left(\frac{3.7\times10^{10}\ \frac{\text{disintegrations}}{\mathrm{s}}}{1\ \mathrm{Ci}}\right)$$

$$\times\left(\frac{1\ \mathrm{Bq}}{1\ \frac{\text{disintegration}}{\mathrm{s}}}\right)\left(\frac{1\ \mathrm{gal}}{3.79\ \mathrm{L}}\right)$$

$$= \boxed{1.95\ \mathrm{Bq/L} > 0.2\ \mathrm{Bq/L}}$$

Therefore, this does violate EPA standards.[47] Thus, A or B is the answer.

A look at the chart of the nuclides reveals that radium is primarily an alpha emitter. This type of radiation causes highly localized damage yet cannot penetrate the outer layers of skin. Therefore, it is an internal hazard.

Answer is A.

[46] This element is used in industrial radiography and for medical purposes.

[47] Familiarity with the more common limits for radiation exposure is useful. Additionally, a knowledge of exposure sources and values is useful (e.g., natural background radiation absorbed dose equivalent is approximately 30 mSv or 300 mRem per year in the US).

NUCLEAR RADIATION–22

Reactor siting is an important aspect of the design and licensing of a nuclear plant. What factors are considered by the NRC in evaluating a site?[48]

(A) reactor design and population density
(B) site physical characteristics
(C) engineering safeguards
(D) all of the above

Solution:

The reactor design, especially the expected inventory of radioactive material, is certainly a factor with new designs requiring additional margins of safety.

A low population density is desirable. The area surrounding a reactor is divided into zones, each of which has its own requirements. Numerous detailed requirements exist for this siting factor.

The site physical characteristics include the seismology (e.g., earthquake concerns); meteorology (e.g., dispersal of radioactive releases); geology (e.g., structural concerns); and hydrology (e.g., safety related effects during flooding), all of which impact siting.

Finally, engineering safeguards. This may seem out of place in terms of siting; consider, however, the safeguards designed into a plant to mitigate the negative affect of one of the above factors. For example, additional structural strength near earthquake faults and so on. Therefore, an engineering safeguard may overcome a negative aspect of the other factors.

Thus, all of the mentioned items are factors in siting.

Answer is D.

Note: Once a reactor is in a plant, environmental monitoring of the soil, air, and water around a plant occurs to ensure compliance with standards and regulations.

[48] See 10CFR100 (i.e., title 10, Code of Federal Regulations, chapter 100) for additional information. The Atomic Energy Act of 1954, court decisions and Regulatory Guides also provide information on reactor siting.

NUCLEAR RADIATION–23

Numerous surface decontamination methods exist. Consider the following.

designation	method/material
I	vacuum cleaning
II	water
III	steam
IV	abrasion

Each method has advantages and disadvantages and must be picked based on the nature of the problem to be resolved. Consider the following.

designation	remarks
a	Activity is reduced to as low as desired. Contamination can spread.
b	Activity is reduced by approximately 90%, and this method can be used on oily surfaces.
c	This method requires the use of a filter.
d	Activity is reduced by approximately 50%.

Match each of the methods with the appropriate remarks.

(A) I a II b III c IV d
(B) I c II d III a IV b
(C) I a II b III d IV c
(D) I c II d III b IV a

Solution:

The use of a vacuum cleaner is effective for dry spills involving loose, not embedded, contamination. To prevent the spread of contamination during the process, a filter must be added. Thus I-c is the correct combination.

Water has been shown to effective in decontamination efforts in that it reduces the original radioactive levels by approximately 50%. Thus II-d is the correct combination.

Steam reduces the activity by approximately 90% and, unlike water, can be used on oily surfaces. Thus III-b is the correct combination.

Abrasive methods are able to reduce the radioactive contamination to any desired level but must be handled correctly to prevent the further spread of contamination and to minimize exposure to personnel. Thus IV-a is the correct combination.

Answer is D.

10 NUCLEAR THEORY: SOLUTIONS

NUCLEAR THEORY-1

Approximately how much energy is released in the conversion/annihilation of 1 lb of mass?

(A) 400 kw/hr
(B) 4000 kw/hr
(C) 11 000 000 kw/hr
(D) 11 000 000 000 kw/hr

Solution:

The conversion of mass to energy is the underlying process in a nuclear reactor. The equivalency between the two was established in Einstein's theory of relativity and is represented by

$$E_0 = m_0 c^2$$

E_0 = rest energy
m_0 = rest mass
c = speed of light

Substituting gives

$$\begin{aligned} E_0 &= m_0 c^2 \\ &= (1\ \text{lb})\left(\frac{1\ \text{kg}}{2.2\ \text{lb}}\right)\left(3.0\times 10^8\ \frac{\text{m}}{\text{s}}\right)^2 \\ &= 4.09\times 10^{16}\ \text{J} \end{aligned}$$

Convert to the desired units as follows.

$$\begin{aligned} E_0 &= 4.09\times 10^{16}\ \text{J} \\ &= \left(4.09\times 10^{16}\ \text{J}\right)\left(\frac{1\ \text{kw-hr}}{3.60\times 10^6\ \text{J}}\right) \\ &= \boxed{11.36\times 10^9\ \text{kw-hr}\quad (11\ \text{billion kw-hr})} \end{aligned}$$

Answer is D.

Note: If Einstein's equation is used to represent the total energy (i.e., including kinetic energy), it takes the form

$$E_{\text{tot}} = m_0 c^2 \left(\frac{1}{1-\frac{v^2}{c^2}} - 1 \right)$$

v = speed

This is the applicable equation for electrons greater than 10 keV and for neutrons greater than 20 MeV.[1]

NUCLEAR THEORY-2

What is the approximate binding energy per nucleon of U-235?

(A) 1.5 MeV
(B) 7.8 MeV
(C) 235 MeV
(D) 1700 MeV

Solution:

The binding energy per nucleon is given by

$$\frac{BE}{A} \cong \frac{\left(Zm_H + (A-Z)m_n - m_x\right)\left(c^2\right)}{A}$$

BE = binding energy
Z = number of protons
A = number of nucleons
m_H = rest mass of a hydrogen atom in amu
m_n = rest mass of a neutron in amu
m_x = rest mass of the element in amu

[1] These energies represent about 20% of the rest mass energy. Practically speaking, the formula is not required for most neutron calculations in nuclear engineering.

Data is obtained from Table 4.1 in App. 4.A.[2]

$$\frac{BE}{A} \cong \frac{(Zm_H + (A-Z)m_n - m_x)(c^2)}{A}$$

$$\cong \frac{\begin{pmatrix}(92)(1.007825\,\text{amu}) \\ +(235-92)(1.00866\,\text{amu}) \\ -(235\,\text{amu})\left(\frac{1.66054\times10^{-24}\,\text{g}}{1\,\text{amu}}\right)\left(\frac{1\,\text{kg}}{1\times10^3\,\text{g}}\right)\end{pmatrix} \times\left(3.00\times10^8\,\frac{\text{m}}{\text{s}}\right)^2}{235}$$

$$\cong (1.25\times10^{-12}\,\text{J})\left(\frac{1\,\text{eV}}{1.60\times10^{-19}\,\text{J}}\right)\left(\frac{1\,\text{MeV}}{1\times10^6\,\text{eV}}\right)$$

$$\cong \boxed{7.78\,\text{MeV}\quad(7.8\,\text{MeV})}$$

Answer is B.

The binding energy per nucleon curve is as shown.

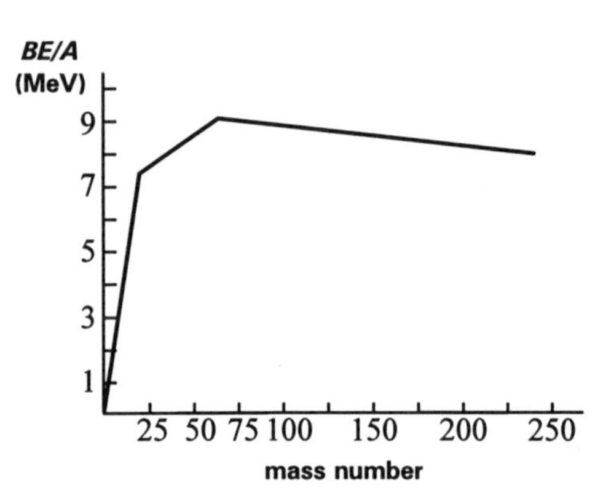

Elements to the right of mass number 60 on this curve fission to release energy. Elements to the left of 60 undergo fusion to release energy.

[2] The answer given is approximate due to using inexact amounts for the mass. The exact mass of U-235 is 235.04 amu. Additionally, some other effects are unaccounted for by this formula. See the Reference section for more information. Note the electron rest masses cancel by using the mass of the hydrogen atom, which contains an electron, and the mass of the element, which also includes electrons. The binding energy of the electrons is not canceled. The effect of all of the above is minor.

NUCLEAR THEORY–3

The energy distribution among gas atoms or molecules tends to follow the Maxwellian distribution given by

$$N(E) = \left(\frac{2\pi N}{(\pi kT)^{3/2}}\right)\left(E^{1/2}\right)\left(e^{-\frac{E}{kT}}\right)$$

$N(E)$ = number of particles with energy E
k = Boltzmann's constant
T = temperature in kelvin

A plot of the energies takes the following shape.

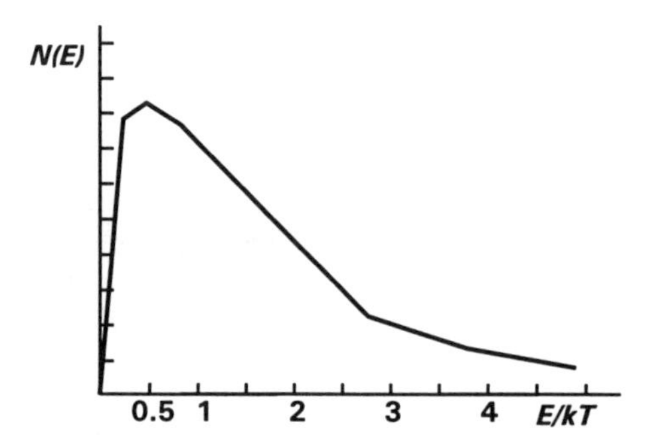

Above room temperature (i.e., approximately 300 K) the equation can be used for liquids and solids with minimal error. The most probable energy is

$$E_p = \frac{1}{2}kT$$

The average energy is

$$\overline{E} = \frac{3}{2}kT$$

Assuming a reference temperature T_0 of 293.61 K (20°C), what is the energy associated with the term kT?[3] What is the speed of a neutron whose energy is given by the term kT?

(A) 0.0253 eV; 2200 m/s
(B) 0.0253 eV; 1560 m/s
(C) 0.511 MeV; 2200 m/s
(D) 1.02 MeV; 1560 m/s

[3] The value of T_0 (20°C) is the temperature to which many nuclear parameters are referenced.

Solution:

Boltzmann's constant is given in Table 4.1 of App 4.A. The energy is

$$\begin{aligned} E_p &= kT_0 \\ &= \left(1.38\times10^{-23}\ \frac{\text{J}}{\text{K}}\right)(293.61\ \text{K}) \\ &= \left(4.05\times10^{-21}\ \text{J}\right)\left(\frac{1\ \text{eV}}{1.602\times10^{-19}\ \text{J}}\right) \\ &= \boxed{0.0253\ \text{eV}} \end{aligned}$$

Therefore, either A or B is the answer.

From the problem statement, equate the kinetic energy of a neutron with the energy given by the term kT. Thus,

$$\begin{aligned} \frac{1}{2}m_n v^2 &= kT_0 \\ v^2 &= \frac{2kT_0}{m_n} \\ v &= \sqrt{\frac{2kT_0}{m_n}} \end{aligned}$$

Substituting data from Table 4.1 of App. 4.A gives

$$\begin{aligned} v &= \sqrt{\frac{2kT_0}{m_n}} \\ &= \sqrt{\frac{(2)\left(1.38\times10^{-23}\ \frac{\text{J}}{\text{K}}\right)(293.61\ \text{K})}{\left(1.67492\times10^{-24}\ \text{g}\right)\left(\frac{1\ \text{kg}}{1\times10^{3}\ \text{g}}\right)}} \\ &= \boxed{2.2\times10^{3}\ \text{m/s}\quad(2200\ \text{m/s})} \end{aligned}$$

Answer is A .

NUCLEAR THEORY–4

A point source emits 10^{13} neutrons/s into a water moderator as shown.

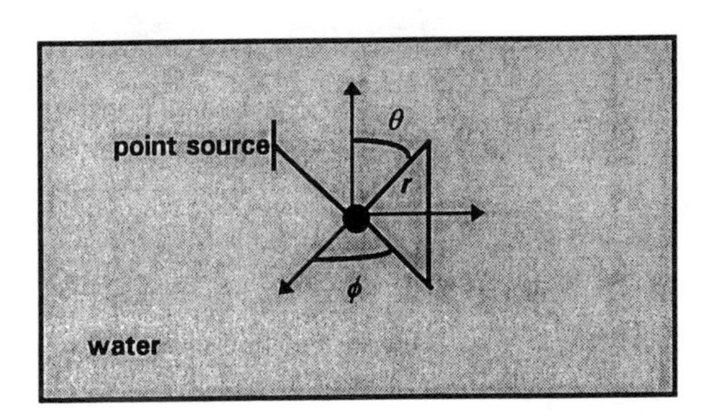

The neutron flux is given by[4]

$$\phi(r) = \frac{Se^{-\frac{r}{L}}}{4\pi Dr}$$

S = source strength neutrons/s
L = diffusion length
D = diffusion coefficient
r = radial position

For water, approximate values for the defined quantities are

$$L = 3\ \text{cm}$$
$$D = 0.3\ \text{cm}$$

What is the net rate of neutrons flowing out of a sphere of radius 2 m surrounding the point source? Use Fick's law (i.e., the diffusion approximation).

(A) $8\times10^{-15}\ \text{s}^{-1}$
(B) $9\times10^{-12}\ \text{s}^{-1}$
(C) $9\times10^{12}\ \text{s}^{-1}$
(D) $8\times10^{15}\ \text{s}^{-1}$

[4] This formula assumes an infinite moderator and uses as its basis neutron diffusion (i.e., Fick's law). See the Reference section for more information.

Solution:

The neutron current density vector, derived from Fick's law (using flux instead of neutron density) is[5]

$$\vec{J} = -D\vec{\nabla}\phi$$

Spherical geometry is clearly called for with the neutron current density vector pointing radially outward. Therefore, only the radial component of the gradient need be considered. Thus,

$$\begin{aligned}\vec{J} &= -D\vec{\nabla}\phi \\ &= -D\left(\vec{n}\frac{d}{dr}\left(\frac{Se^{-\frac{r}{L}}}{4\pi Dr}\right)\right) \\ &= -D\vec{n}\left(\frac{S}{4\pi D}\frac{d}{dr}\left(\frac{e^{-\frac{r}{L}}}{r}\right)\right) \\ &= -\vec{n}\frac{S}{4\pi}\left(e^{-\frac{r}{L}}\left(\frac{-1}{r^2}\right) + \frac{e^{-\frac{r}{L}}\left(\frac{-1}{L}\right)}{r}\right) \\ &= \vec{n}\frac{S}{4\pi}e^{-\frac{r}{L}}\left(\frac{1}{r^2}+\frac{1}{rL}\right)\end{aligned}$$

The question referred to the net number of neutrons flowing which is given by the following.

$$\vec{J}\bullet\vec{n} = J_r$$

For the net number of neutrons over the entire sphere the value of J_r must be multiplied by the area.

$$\text{net neutrons} = J_r 4\pi r^2$$

Therefore the desired quantity can be obtained from the following.

$$\begin{aligned}J_r 4\pi r^2 &= \left|\vec{J}(r)\right|\left(4\pi r^2\right) \\ &= \left(\frac{S}{4\pi}\right)\left(e^{-\frac{r}{L}}\right)\left(\frac{1}{r^2}+\frac{1}{rL}\right)\left(4\pi r^2\right) \\ &= Se^{-\frac{r}{L}}\left(1+\frac{r}{L}\right)\end{aligned}$$

Substituting the given values gives

$$\begin{aligned}J_r 4\pi r^2 &= Se^{-\frac{r}{L}}\left(1+\frac{r}{L}\right) \\ &= \left(1\times10^{13}\ \frac{\text{neutrons}}{\text{s}}\right)\exp\left(-\frac{(2\ \text{m})\left(\frac{10^2\ \text{cm}}{1\ \text{m}}\right)}{3\ \text{cm}}\right) \\ &\quad\times\left(1+\frac{(2\ \text{m})\left(\frac{10^2\ \text{cm}}{1\ \text{m}}\right)}{3\ \text{cm}}\right) \\ &= \boxed{7.54\times10^{-15}\ \text{neutrons/s}\quad\left(8\times10^{-15}\ \text{s}^{-1}\right)}\end{aligned}$$

Note: The number is small as expected for a small diffusion length and large sphere of water.[6] The important concepts are the use of Fick's law in determining the neutron current density vector and the use of the gradient in the appropriate coordinate system.[7]

Answer is A .

[5] This assumes neutrons diffuse much like gas atoms or molecules. The use of this diffusion approximation results in a diffusion equation from which the neutron flux can be determined, both spatially and over time. More accurate methods exist but the results obtained from assuming diffusion are good first estimates of reactor properties.

[6] Water is an excellent neutron shield. See the Nuclear Radiation section for additional information.

[7] Fick's law is not valid if the medium strongly absorbs neutrons, within three mean free paths of a source or surface, or where the scattering of neutrons is strongly anisotropic.

NUCLEAR THEORY–5

The following equation represents the general form of the neutron continuity equation.

$$\frac{\delta n}{\delta t} = S - \Sigma_a \phi - \text{div}\vec{J}$$

Which term represents leakage of neutrons from the volume under consideration?

(A) S
(B) $\Sigma_a\phi$
(C) div$\boldsymbol{J}$
(D) B or C

Solution:

The neutron continuity equation in word terms is

$$\begin{pmatrix}\text{rate of change}\\ \text{of neutrons}\end{pmatrix} = \begin{pmatrix}\text{rate of production}\\ \text{of neutrons}\end{pmatrix} - \begin{pmatrix}\text{rate of absorption}\\ \text{of neutrons}\end{pmatrix} - \begin{pmatrix}\text{rate of leakage}\\ \text{of neutrons}\end{pmatrix}$$

In mathematical terms, using rectangular coordinates, the final term (i.e., the rate of leakage) is

$$\text{div}\vec{J} = \frac{\delta J_x}{\delta x} + \frac{\delta J_y}{\delta y} + \frac{\delta J_z}{\delta z}$$

Answer is C.

Note: Substituting Fick's law into the neutron continuity equation gives

$$\begin{aligned}\frac{\delta n}{\delta t} &= S - \Sigma_a \phi - \text{div}\vec{J}\\ &= S - \Sigma_a \phi - \text{div}\left(-D\vec{\nabla}\phi\right)\end{aligned}$$

Assuming the diffusion coefficient D is independent of position gives

$$\frac{\delta n}{\delta t} = S - \Sigma_a \phi + D\nabla^2 \phi$$

Rearranging and assuming steady state gives

$$D\nabla^2 \phi - \Sigma_a \phi + S = 0$$

Divide through by the diffusion coefficient D.

$$\nabla^2 \phi - \frac{\Sigma_a}{D}\phi + \frac{S}{D} = 0$$

Rearrange and define the diffusion area L^2 as D/Σ_a.

$$\nabla^2 \phi - \frac{1}{L^2}\phi = -\frac{S}{D}$$

This is the steady state diffusion equation and is used to determine the neutron flux for various geometries (e.g., the point source flux used in Prob. 4).

NUCLEAR THEORY–6

The diffusion equation previously used assumed monoenergetic neutrons (i.e., it can be considered a one-group equation). For thermal reactors at least two groups of neutron energies should be used, fast and thermal.[8]

The energy groups are organized as follows.

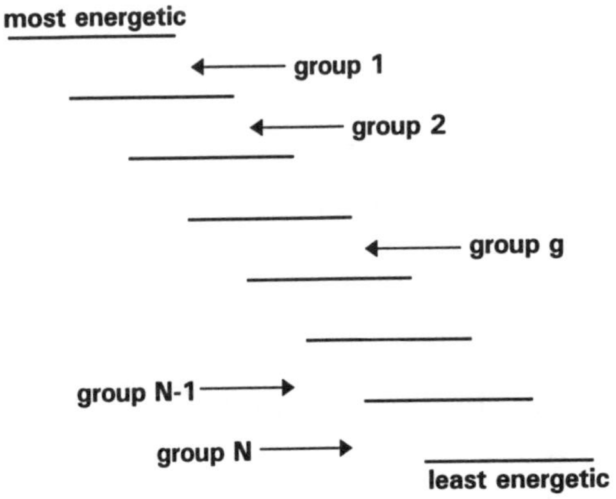

most energetic
group 1
group 2
group g
group N-1
group N
least energetic

[8] These calculations can involve many neutron energy groups and are termed multi-group calculations which normally involve the use of computers to aid in determination of the fluxes. This is a much more exact method of determining reactor properties than using one- and two-group diffusion results.

Consider a certain fast reactor, fueled with U-238 and moderated by Na-23. The neutrons are separated into six groups with the corresponding cross sections shown.[9]

group	σ_{tr} barns	σ_r barns	$\sigma_{g\to g+1}$ barns	ϕ_g cm-3·s-1
1	6.3	0.04	2.44	1×10^{13}
2	7.5	0.07	1.02	3.5×10^{13}
3	0.11	0.10	0.95	4×10^{14}
4	0.11	0.15	0.63	7×10^{14}
5	0.14	0.23	0.40	1×10^{15}
6	0.16	0.61	–	5×10^{16}

Sodium density is 0.97 g/cm³ for Na-11.[10]

What is most nearly the rate at which neutrons are scattered from the second into the third group?

(A) 6×10^{11} cm-3·s-1
(B) 9×10^{11} cm-3·s-1
(C) 10×10^{12} cm-3·s-1
(D) 9×10^{13} cm-3·s-1

Solution:

Let the scattering rate be F. Thus, the scattering rate is

$$
\begin{aligned}
F_{2\to 3} &= \Sigma_{2\to 3}\phi_2 \\
&= N\sigma_{2\to 3}\phi_2 \\
&= \left(\frac{\rho N_A}{MW}\right)\sigma_{2\to 3}\phi_2 \\
&= \left(\frac{\left(0.97\ \frac{\text{g}}{\text{cm}^3}\right)\left(0.602\times 10^{24}\ \frac{\text{atoms}}{\text{mol}}\right)}{23\ \frac{\text{g}}{\text{mol}}}\right) \\
&\quad \times\left(1.02\times 10^{-24}\ \text{cm}^2\right)\left(3.5\times 10^{13}\ \frac{\text{neutrons}}{\text{cm}^2\cdot\text{s}}\right) \\
&= \boxed{9.06\times 10^{11}\ \text{neutrons/cm}^3\cdot\text{s}}
\end{aligned}
$$

Answer is B.

[9] The values given are approximate. Additionally, only g→g+1 cross sections are shown. Cross sections for g→g+2, g→g+3, and so on could be used. Scattering from a lower energy group into a higher energy group (e.g., g+3→g+2) must also be accounted for in complete calculations. Final note, the fluxes are arbitrary.

[10] If the term normal density is used it refers to the atom density × 10^{-24} (i.e., atoms/cm3) for the material of interest at standard conditions.

Note: Multi-group calculations and the methods developed in this section (i.e., Nuclear Theory) can be used to determine the properties for heterogeneous reactors. The difference is that the factors (e.g., $k_\infty = n_{th} f \varepsilon p$) must be determined carefully to account for the various portions of the core.

NUCLEAR THEORY–7

What is the thermal flux corresponding to a neutron density of 10^{13} neutrons/cm³ at 293.61 K?

(A) 2.2×10^{13} cm-3·s-1
(B) 2.5×10^{15} cm-3·s-1
(C) 2.2×10^{18} cm-3·s-1
(D) 2.5×10^{18} cm-3·s-1

Solution:

The thermal flux is given by

$$\phi_{\text{th}} = \frac{2}{\sqrt{\pi}} n v_{\text{th}}$$

This flux ϕ_{th} differs from the 2200 m/s flux ϕ_0 in that it is calculated from[11]

$$\phi_T = \int_{\text{th}} \phi(E)dE$$

By contrast, the flux ϕ_0 is calculated assuming monoenergetic neutrons of energy $E_0 = 0.0253$ eV.

Using data from Table 4.1 in App. 4.A, v_{th} is

$$
\begin{aligned}
E_{\text{th}} &= kT = \frac{1}{2} m v_{\text{th}}^2 \\
v_{\text{th}} &= \sqrt{\frac{2kT}{m}} \\
&= \sqrt{\frac{(2)\left(1.38\times 10^{-23}\ \frac{\text{J}}{\text{K}}\right)(293.61\ \text{K})}{\left(1.67492\times 10^{-24}\ \text{g}\right)\left(\frac{1\ \text{kg}}{10^3\ \text{kg}}\right)}} \\
&= 2200\ \text{m/s}
\end{aligned}
$$

[11] The subscript "th" indicates the integration is carried out from zero to the top of the thermal energy band (i.e., approximately $5kT$). See the Reference section for more information.

Note: The thermal velocity equals 2200 m/s only because of the chosen temperature.

Substituting,

$$\begin{aligned}\phi_{\text{th}} &= \frac{2}{\sqrt{\pi}} n v_{\text{th}} \\ &= \left(\frac{2}{\sqrt{\pi}}\right)\left(10^{13}\ \frac{\text{neutrons}}{\text{cm}^3}\right)\left(2200\ \frac{\text{m}}{\text{s}}\right)\left(\frac{10^2\ \text{cm}}{1\ \text{m}}\right) \\ &= \boxed{2.48\times10^{18}\ \text{cm}^{-2}\cdot\text{s}^{-1}\quad \left(2.5\times10^{18}\ \text{cm}^{-2}\cdot\text{s}^{-1}\right)}\end{aligned}$$

Answer is D .

NUCLEAR THEORY–8

What is the ratio of the 2200 m/s flux to the thermal flux given an operating temperature of 597 K?

(A) 0.5
(B) 0.6
(C) 0.7
(D) 0.8

Solution:

The 2200 m/s flux is given by

$$\phi_0 = n v_0$$

The thermal flux is given by (see Prob. 7)

$$\phi_{\text{th}} = \frac{2}{\sqrt{\pi}} n v_{\text{th}}$$

The ratio is then given by

$$\frac{\phi_0}{\phi_{\text{th}}} = \frac{n v_0}{\frac{2}{\sqrt{\pi}} n v_T} = \left(\frac{\sqrt{\pi}}{2}\right)\left(\frac{v_0}{v_T}\right)$$

To get the temperature into the relationship use the following.

$$E_{\text{th}} = \frac{1}{2} m v_T^{\,2} = kT$$

Rearranging gives

$$v_T = \sqrt{\frac{2kT}{m}}$$

For the 2200 meter flux use the following.

$$E_0 = \frac{1}{2} m v_0^{\,2} = kT_0$$

Rearranging gives

$$v_0 = \sqrt{\frac{2kT_0}{m}}$$

Substituting these terms into the ratio equation gives the desired result.

$$\begin{aligned}\frac{\phi_0}{\phi_{\text{th}}} &= \left(\frac{\sqrt{\pi}}{2}\right)\left(\frac{v_0}{v_{\text{th}}}\right) \\ &= \left(\frac{\sqrt{\pi}}{2}\right)\left(\frac{\sqrt{\frac{2kT_0}{m}}}{\sqrt{\frac{2kT}{m}}}\right) \\ &= \left(\frac{\sqrt{\pi}}{2}\right)\left(\sqrt{\frac{T_0}{T}}\right)\end{aligned}$$

Substitute the given values.

$$\begin{aligned}\frac{\phi_0}{\phi_{\text{th}}} &= \left(\frac{\sqrt{\pi}}{2}\right)\left(\sqrt{\frac{T_0}{T}}\right) \\ &= \left(\frac{\sqrt{\pi}}{2}\right)\left(\sqrt{\frac{293.61\ \text{K}}{597\ \text{K}}}\right) \\ &= \boxed{0.622\quad (0.6)}\end{aligned}$$

Answer is B .

Note: The important concept here is that the thermal flux ϕ_{th} is the flux to use in calculations determining the diffusion of neutrons in an operating reactor. The 2200 m/s flux is appropriate for calculations when determining neutron absorption rates in target nuclei.

NUCLEAR THEORY–9

The one-group diffusion equation is[12]

$$D\nabla^2\phi - \Sigma_a\phi + S = 0 \text{ or}$$
$$D\nabla^2\phi - \Sigma_a\phi = -S$$

The source term S is due to fissions.[13] Thus,

$$S = \eta\Sigma_{a,\text{fuel}}\phi$$
$$= \eta\left(\frac{\Sigma_{a,\text{fuel}}}{\Sigma_a}\right)\Sigma_a\phi$$
$$= \eta f\Sigma_a\phi$$

η = average number of fission neutrons emitted per neutron absorbed in the fuel, called the regeneration factor

$$\eta = \nu\left(\frac{\sigma_f}{\sigma_a}\right)$$

ν = average number of neutrons emitted during fission, both prompt and delayed

$\Sigma_{a,\text{fuel}}$ = macroscopic absorption cross section in the fuel

Σ_a = macroscopic absorption cross section in all material

f = fuel utilization[14]

The infinite multiplication factor is the number of neutrons in one generation divided by the number of neutrons in the preceding generation.

$$k_\infty = \frac{\eta f\Sigma_a\phi}{\Sigma_a\phi} = \eta f$$

Substituting into the diffusion equations gives

$$D\nabla^2\phi - \Sigma_a\phi = -k_\infty\Sigma_a\phi$$
$$D\nabla^2\phi + (k_\infty - 1)\Sigma_a\phi = 0$$
$$\nabla^2\phi + (k_\infty - 1)\frac{\Sigma_a}{D}\phi = 0$$
$$\nabla^2\phi + \frac{(k_\infty - 1)}{L^2}\phi = 0$$

The term $L^2 = D/\Sigma_a$ and is called the diffusion area.

Define the *buckling* B^2 as

$$B^2 = \frac{k_\infty - 1}{L^2}$$

Substituting gives the final result.

$$\nabla^2\phi + B^2\phi = 0$$

This is called the one-group reactor equation. The buckling is determined by the material properties of the system only.[15]

What is the value for the fuel utilization f and the infinite multiplication factor k_∞ for a homogeneous mixture of U-235 and iron in which the uranium is 1 w/o (i.e., weight percent) of the total?

(A) f 0.394 k_∞ 0.867
(B) f 0.654 k_∞ 1.43
(C) f 0.922 k_∞ 2.03
(D) f 11.9 k_∞ 26.1

Solution:

The fuel utilization is given by

$$f = \frac{\Sigma_{a,\text{fuel}}}{\Sigma_a}$$

[12] The explanation and equations that follow are applicable to one-group (i.e., fast) reactors. The principles can be extended to thermal reactors.

[13] This is true for a critical reactor. Several sources of neutrons exist for a subcritical reactor with a minor number of them coming from spontaneous fission.

[14] This term is called the thermal utilization factor in thermal reactors.

[15] The diffusion equation which started this derivation assumes a bare reactor (i.e., a reactor without a blanket or a reflector). See the Reference section for a text defining nuclear terms.

Expand the macroscopic cross sections and rearrange to obtain a ratio of the atom densities.

$$f = \frac{\Sigma_{a,\text{fuel}}}{\Sigma_a}$$

$$= \frac{N_{\text{fuel}}\sigma_{a,\text{fuel}}}{N_{\text{fuel}}\sigma_{a,\text{fuel}} + N_{Fe}\sigma_{a,Fe}}$$

$$= \frac{1}{1+\frac{N_{Fe}\sigma_{a,Fe}}{N_{\text{fuel}}\sigma_{a,\text{fuel}}}}$$

Since the atom densities were not provided, the weight percent must be used. Let ρ_{Fe} and ρ_F be the density of the iron and the fuel respectively. Thus,

$$\rho_{Fe} = \frac{N_{Fe}AW_{Fe}}{N_A}$$

$$\rho_F = \frac{N_F AW_F}{N_A}$$

$$\frac{\rho_{Fe}}{\rho_F} = \frac{N_{Fe}AW_{Fe}}{N_F AW_F}$$

Use the above information to solve for the atom density ratio.

$$\frac{N_{Fe}}{N_F} = \frac{\rho_{Fe}AW_F}{\rho_F AW_{Fe}}$$

AW = gram atomic weight

The fuel is 1 w/o (i.e., weight percent) of the total. Therefore,

$$\frac{\rho_F}{\rho_F+\rho_{Fe}} = 0.01$$

$$\rho_F = (\rho_F+\rho_{Fe})(0.01)$$

$$100\rho_F = \rho_F+\rho_{Fe}$$

$$99\rho_F = \rho_{Fe}$$

$$99 = \frac{\rho_{Fe}}{\rho_F}$$

Substituting this information into the density ratio formula gives the following.

$$\frac{N_{Fe}}{N_F} = \frac{\rho_{Fe}AW_F}{\rho_F AW_{Fe}}$$

$$\frac{N_{Fe}}{N_F} = (99)\left(\frac{235\,\frac{\text{g}}{\text{mol}}}{55\,\frac{\text{g}}{\text{mol}}}\right)$$

$$= 423$$

Substitute this value for the density ratio and the values for the microscopic absorption cross sections from Table 4.2 in App. 4.A.

$$f = \frac{1}{1+\frac{N_{Fel}\sigma_{a,Fe}}{N_{\text{fuel}}\sigma_{a,\text{fuel}}}}$$

$$= \frac{1}{1+(423)\left(\frac{0.006\ \text{barns}}{1.65\ \text{barns}}\right)}$$

$$= \boxed{0.394}$$

The multiplication factor, for an infinite core, is

$$k_\infty = \eta f$$

Substitute the calculated value for the fuel utilization and the information from Table 4.2 App. 4.A.

$$k_\infty = \eta f$$

$$= (2.2)(0.394)$$

$$= \boxed{0.867}$$

An infinite core with this composition would be subcritical since the multiplication factor is less than one.

Answer is A.

Nuclear Theory (NT) problems 10 and 11 are based on the following information and equations.

A spherical reactor shape is often used as a first estimate of expected properties. Recall, the one-group diffusion equation is

$$\nabla^2\phi + B^2\phi = 0$$

Using this equation and solving for the flux in a sphere of radius R gives the following.

buckling	flux	constant A
$\left(\frac{\pi}{R}\right)^2$	$A\left(\frac{1}{r}\right)\sin\left(\frac{\pi r}{R}\right)$	$\frac{P}{4R^2 E_R \Sigma_f}$

E_R = recoverable energy per fission in Joules per fission
Σ_f = macroscopic fission cross section
P = operating power (which determines the magnitude of the flux)
$(\pi/R)^2$ = the first eigenvalue squared $(B_1)^2$ which is the solution to the flux equation[16]

Since the value for buckling must equal the first eigenvalue the following holds true.

$$\frac{k_\infty - 1}{L^2} = B_1^2$$

Note: The important concept here is that the left side of this equation is determined by the material properties of the system. The right hand side of the equation is determined by the geometry and dimensions. The two together determine the requirements for criticality.

The subscript indicating the first eigenvalue is usually dropped. Rearranging the equation into the form most often seen gives the following.

$$\frac{k_\infty}{1 + L^2 B^2} = 1$$

This is the one-group critical equation for a bare reactor.

[16] The one-group diffusion equation is solved to determine the necessary flux equation (i.e., the flux shape). The buckling must equal the first eigenvalue squared in order to meet boundary conditions. The first eigenvalue is the only one of concern in a critical reactor. See the Reference section texts for a more complete explanation.

miscellaneous information

item	name	value
N_{fe}	atom density	0.848×10^{24} atoms/cm³
N_{pu}	atom density	0.0493×10^{24} atoms/cm³
L^2	diffusion area	D/Σ_a

NUCLEAR THEORY–10

A homogeneous mixture of Fe-55 and Pu-239 is to be made into the shape of a bare sphere. Estimate the critical radius of this fast reactor?

(A) 2 cm
(B) 6 cm
(C) 12 cm
(D) 66 cm

Solution:

The one-group critical equation is

$$\frac{k_\infty}{1 + L^2 B^2} = 1$$

The buckling for a sphere is

$$B^2 = \left(\frac{\pi}{R}\right)^2$$

Substituting gives

$$\frac{k_\infty}{1 + L^2\left(\frac{\pi}{R}\right)^2} = 1$$

Solve for R.

$$\frac{k_\infty}{1 + L^2\left(\frac{\pi}{R}\right)^2} = 1$$

$$k_\infty = 1 + L^2\left(\frac{\pi}{R}\right)^2$$

$$\frac{k_\infty - 1}{L^2} = \left(\frac{\pi}{R}\right)^2$$

$$R^2 = \frac{L^2\pi^2}{k_\infty - 1}$$

$$R = \sqrt{\frac{L^2\pi^2}{k_\infty - 1}}$$

$$= \pi\sqrt{\frac{L^2}{k_\infty - 1}}$$

Determine the diffusion area.

$$L^2 = \frac{D}{\Sigma_a}$$

The diffusion coefficient D can be estimated from the following.

$$D \cong \frac{\tau_{tr}}{3}$$

τ_{tr} = transport mean free path ($1/\Sigma_{tr}$)

Therefore,

$$D \cong \frac{\tau_{tr}}{3}$$

$$\cong \left(\frac{1}{3}\right)\left(\frac{1}{\Sigma_{tr}}\right) = \frac{1}{3\Sigma_{tr}}$$

From the information given in Table 4.2 of App. 4.A,

$$\Sigma_{tr} = N_{Pu}\sigma_{tr,Pu} + N_{Fe}\sigma_{tr,Fe}$$

$$= \left(0.0493 \times 10^{24}\ \text{cm}^{-3}\right)\left(6.8\ \text{barns}\right)\left(\frac{10^{-24}\ \text{cm}^2}{1\ \text{barn}}\right)$$

$$+ \left(0.0848 \times 10^{24}\ \text{cm}^{-3}\right)\left(2.7\ \text{barns}\right)\left(\frac{10^{-24}\ \text{cm}^2}{1\ \text{barn}}\right)$$

$$= 0.564\ \text{cm}^{-1}$$

Thus,

$$D \cong \frac{1}{3\Sigma_{tr}}$$

$$\cong \frac{1}{(3)\left(0.564\ \text{cm}^{-1}\right)}$$

$$\cong 0.591\ \text{cm}$$

The total macroscopic cross section for absorption in the mixture can be determined from the summation of the individual macroscopic cross sections. Thus,

$$\Sigma_a = \Sigma_{a,Pu} + \Sigma_{a,Fe}$$

$$= N_{Pu}\sigma_{a,Pu} + N_{Fe}\sigma_{a,Fe}$$

$$= \left(0.0493 \times 10^{24}\ \text{cm}^{-3}\right)\left(2.11\ \text{barns}\right)\left(\frac{10^{-24}\ \text{cm}^2}{1\ \text{barn}}\right)$$

$$+ \left(0.0848 \times 10^{24}\ \text{cm}^{-3}\right)\left(0.006\ \text{barns}\right)\left(\frac{10^{-24}\ \text{cm}^2}{1\ \text{barn}}\right)$$

$$= 0.105\ \text{cm}^{-1}$$

Substitute the calculated values into the equation for the diffusion area.

$$L^2 = \frac{D}{\Sigma_a}$$

$$= \frac{0.591\ \text{cm}}{0.105\ \text{cm}^{-1}}$$

$$= 5.63\ \text{cm}^2$$

Determine the other unknown in the critical radius equation, k_∞.

$$k_\infty = \eta f$$

The fuel utilization can be determined from the following.

$$f = \frac{\Sigma_{a,\text{fuel}}}{\Sigma_a}$$

$$= \frac{\Sigma_{a,Pu}}{\Sigma_{a,Pu} + \Sigma_{a,Fe}}$$

From the previous calculation,

$$\Sigma_{a,Pu} = N_{Pu}\sigma_{Pu} = 0.104\ \text{cm}^{-1}$$

$$\Sigma_{a,Fe} = N_{Fe}\sigma_{Fe} = 0.00051\ \text{cm}^{-1}$$

$$\Sigma_a = 0.105\ \text{cm}^{-1}$$

Substituting gives

$$\begin{aligned} f &= \frac{\Sigma_{a,Pu}}{\Sigma_{a,Pu} + \Sigma_{a,Fe}} \\ &= \frac{0.104\ \text{cm}^{-1}}{0.104\ \text{cm}^{-1} + 0.00051\ \text{cm}^{-1}} \\ &= 0.99 \end{aligned}$$

The value of η in a mixture is given by the following.

$$\begin{aligned} \eta &= \frac{1}{\Sigma_a} \sum_i \nu(i)\Sigma_f(i) \\ &= \frac{\nu(\text{Pu-239})\Sigma_f(\text{Pu-239})}{\Sigma_a(\text{Pu-239}) + \Sigma_a(\text{Fe-55})} \end{aligned}$$

Using values previously calculated and data from Table 4.2 of App. 4.A gives

$$\begin{aligned} \eta &= \frac{\nu_{Pu}\Sigma_{f,Pu}}{\Sigma_{a,Pu} + \Sigma_{a,Fe}} = \frac{\nu_{Pu} N_{Pu} \sigma_{f,Pu}}{\Sigma_{a,Pu} + \Sigma_{a,Fe}} \\ &= \frac{(3.0)(0.0493 \times 10^{24}\ \text{cm}^{-3})(1.85\ \text{barns})\left(\frac{10^{-24}\ \text{cm}^2}{1\ \text{barn}}\right)}{0.104\ \text{cm}^{-1} + 0.00051\ \text{cm}^{-1}} \\ &= 2.61 \end{aligned}$$

Therefore,

$$\begin{aligned} k_\infty &= \eta f \\ &= (2.61)(0.99) \\ &= 2.58 \end{aligned}$$

Substitute the values for the diffusion area L^2 and the infinite multiplication factor k_∞ into the critical radius equation.

$$\begin{aligned} R &= \pi\sqrt{\frac{L^2}{k_\infty - 1}} \\ &= \pi\sqrt{\frac{5.63\ \text{cm}^2}{2.58 - 1}} \\ &= \boxed{5.93\ \text{cm}\quad (6\ \text{cm})} \end{aligned}$$

Answer is B.

NUCLEAR THEORY–11

What is the non-leakage probability for the bare reactor in the Prob. 10?

(A) 0.1
(B) 0.3
(C) 0.4
(D) 0.6

Solution:

The one-group equation for the non-leakage probability for a bare reactor is as follows.

$$P_{NL} = \frac{1}{1 + B^2 L^2}$$

From Prob. 10, the diffusion area L^2 was 5.63 cm^2. For a bare spherical reactor, the buckling is

$$\begin{aligned} B^2 &= \left(\frac{\pi}{R}\right)^2 \\ &= \left(\frac{\pi}{5.93\ \text{cm}}\right)^2 \\ &= 0.281\ \text{cm}^{-2} \end{aligned}$$

Substituting gives

$$\begin{aligned} P_{NL} &= \frac{1}{1 + B^2 L^2} \\ &= \frac{1}{1 + (0.281\ \text{cm}^{-2})(5.63\ \text{cm}^2)} \\ &= \boxed{0.387\quad (0.4)} \end{aligned}$$

Another method for obtaining the solution is to realize that the one-group critical equation contains the non-leakage term.[17]

$$\frac{k_\infty}{1 + B^2 L^2} = 1$$

[17] In fact, this equation represents the probability that neutrons will not leak out multiplied by the multiplication factor. Further, if this value equals one, the reactor is critical (i.e., a self-sustaining chain reaction exists).

Rewriting the critical equation in terms of the non-leakage factor gives the following.

$$k_\infty P_{NL} = 1$$

The multiplication factor was 2.58; therefore,

$$\begin{aligned} k_\infty P_{NL} &= 1 \\ P_{NL} &= \frac{1}{k_\infty} \\ &= \frac{1}{2.58} \\ &= \boxed{0.388 \quad (0.4)} \end{aligned}$$

Answer is C .

NUCLEAR THEORY–12

What is the infinite multiplication factor for a thermal reactor with a thermal utilization of 0.5? Use typical parameter values.

(A) 1.0
(B) 1.5
(C) 2.0
(D) 2.5

Solution:

The infinite multiplication factor k_∞ for a thermal reactor is given by the four factor formula.

$$k_\infty = \eta_{\text{th}} f p \varepsilon$$

Using data from Table 4.3 in App. 4.A gives

$$\begin{aligned} k_\infty &= \eta_{\text{th}} f p \varepsilon \\ &= (2.065)(0.5)(0.95)(1.05) \\ &= \boxed{1.03 \quad (1.0)} \end{aligned}$$

Note: Typical values on η_{th} range from greater than 0.3 to 2.06 for U-235. The thermal utilization is significantly lower than the fuel utilization. The resonance escape probability p is an important parameter in thermal reactor design and can be calculated from the *resonance integral* and the *lethargy*.[18] The fast fission factor ε ranges from 1.02–1.08 for natural or slightly enriched uranium. Importantly, the combination of the resonance escape probability and the fast fission factor is approximately equal to one in many cases (i.e., $p\varepsilon \cong 1$). Further, for a homogeneous mixture of fissile isotopes and moderator only, $p = \varepsilon = 1$ since no resonance absorber for fast neutron induced fission exists.

Answer is A.

NUCLEAR THEORY–13

For a spherical thermal reactor with a water moderator in a homogeneous mixture, what is the value of the infinite multiplication factor necessary for the reactor to be critical? The radius of the reactor is 1 m. Use the modified one-group critical equation.

(A) 1.0
(B) 1.5
(C) 2.0
(D) 2.5

Solution:

The two-group (i.e., fast and thermal) critical equation for a bare reactor which is also known as the modified one-group equation due to similarity in form and terms is as follows.

$$\frac{k_\infty}{1 + B^2 M_T^2} = 1$$

M_T^2= thermal migration area

$$M_T^2 = L_T^2 + \tau_T$$

L_T^2 = thermal diffusion area
τ_T = neutron age[19]

Using values from Table 4.4 of App. 4.B. Recall that the buckling for a spherical reactor is

$$B^2 = \left(\frac{\pi}{R}\right)^2$$

[18] See the Reference section for additional information.
[19] The neutron age is associated with the slowing down of fast neutrons.

Substitute into the modified one-group critical equation.

$$\frac{k_\infty}{1+B^2 M_T{}^2}=1$$

$$\begin{aligned}k_\infty &= 1+B^2 M_T^2\\ &= 1+B^2\left(L_T^2+\tau_T\right)\\ &= 1+\left(\frac{\pi}{R}\right)^2\left(L_T^2+\tau_T\right)\\ &= 1+\left(\frac{\pi}{1\,\text{m}}\right)^2\left(8.1\,\text{cm}^2+27\,\text{cm}^2\right)\left(\frac{1\,\text{m}}{10^2\,\text{cm}}\right)^2\\ &= \boxed{1.03\quad(1.0)}\end{aligned}$$

Answer is A.[20]

Nuclear Theory (NT) problems 14 and 15 are based on the following information and equations.

parameters of a critical bare cylindrical reactor	
buckling	flux
$\left(\frac{2.405}{R}\right)^2+\left(\frac{\pi}{H}\right)^2$	$AJ_0\left(\frac{2.405r}{R}\right)\cos\left(\frac{\pi z}{H}\right)$
constant A	
$\frac{3.63P}{VE_R\Sigma_f}$	

R = radius of the cylinder
r = position from axis center
H = height of core
z = position along the axis ($z = 0$ at $H = 0.5$)
J_0 = Bessel function of the first kind of order zero
P = power
V = volume of core
E_R = recoverable energy per fission (200 MeV)
Σ_f = macroscopic fission cross section

When the thermal flux is the desired quantity, adjustments to the given formulas must be made.[21]

A certain bare cylindrical reactor with a radius of 0.20 m and a height of 1 m consists of a homogeneous mixture of U-235 and ordinary water with a density of 1 g/cm^3.

Note: Ignore density and temperature corrections (i.e., use 20°C values).[22] The non-$1/v$ factor for water may be taken as equal to one.

NUCLEAR THEORY–14

What is the *critical* mass? Use modified one-group theory.

(A) 150 g
(B) 240 g
(C) 1550 g
(D) 4800 g

Solution:

The critical equation from modified one-group theory is

$$\frac{k_\infty}{1+B^2 M_T^2}=1$$

The critical mass is embedded in the multiplication factor term.[23]

[20] The value for k_∞ was calculated differently than in a previous problem since this problem specified a geometry, allowing the buckling B^2 to be calculated. Further, the composition densities weren't specified; thus, individual formulas for η_{TH}, f, p, and ε could not be used.

[21] When the thermal flux is the desired quantity, instead of the 2200 m/s flux, the average macroscopic fission cross section must be used.

$$\overline{\Sigma}_f = N\overline{\sigma}_f = N\left(\frac{\sqrt{\pi}}{2}\right)\left(g_f(t)\right)\left(\sigma_f\left(E_0\right)\right)\left(\sqrt{\frac{T_0}{T}}\right)$$

This includes a term which accounts for the ratio of the 2200 m/s flux ϕ_0 to the thermal flux ϕ_{th}; a non-$1/v$ factor; the 2200 m/s microscopic cross section; and a temperature adjustment if not at 20°C (i.e., 293.61 K). All the macroscopic cross sections should be handled in this manner when used with the thermal flux.

[22] This assumption is made primarily for simplicity. Also, this problem could represent uranium waste mixed with water in a barrel. Thus, the actual temperature is approximately the reference temperature of 20°C (i.e., 293.61 K). Normally the thermal diffusion parameters, the thermal diffusion coefficient and the thermal diffusion area, are adjusted for the density and temperature of the mixture. See the Reference section for additional information.

[23] Recall that for a homogeneous mixture of fissile isotope and moderator, no resonance absorbers exist and no material which results in fast fission exists; therefore, $p = \varepsilon = 1$ and is not shown.

Recall,

$$k_\infty = \eta_{\text{th}} f$$

Specifically, the mass of the fuel necessary to satisfy this equation is in the thermal utilization term f.

Rearranging the original equation gives

$$\frac{k_\infty}{1+B^2 M_T^2} = 1$$
$$k_\infty = 1 + B^2 M_T^2$$

The buckling for a cylindrical reactor is

$$\begin{aligned} B^2 &= \left(\frac{2.405}{R}\right)^2 + \left(\frac{\pi}{H}\right)^2 \\ &= \left(\frac{2.405}{0.20\text{ m}}\right)^2 + \left(\frac{\pi}{1\text{ m}}\right)^2 \\ &= \left(154.5\text{ m}^{-2}\right)\left(\frac{1\text{ m}}{10^2\text{ cm}}\right)^2 \\ &= 1.54\times10^{-2}\text{ cm}^{-2} \end{aligned}$$

The thermal migration area of a mixture may be calculated as follows.

$$M_{\text{mix}}^2 = L_{\text{mix}}^2 + \tau_{\text{mix}}$$

The values for the neutron age and the thermal diffusion coefficient, which is the numerator of the thermal diffusion area, depend primarily on scattering in the medium. As the amount of fissile material required for criticality in a homogenous mixture is small, the properties of the moderator alone can be used. Thus,

$$M_{\text{mix}}^2 \cong M_{T,\text{H2O}}^2 = L_{T,\text{H2O}}^2 + \tau_{T,\text{H2O}}$$

Substitute values from Table 4.4 in App. 4.B.

$$\begin{aligned} M_{\text{mix}}^2 &\cong L_{T,\text{H2O}}^2 + \tau_{T,\text{H2O}} \\ &\cong 8.1\text{ cm}^2 + 27\text{ cm}^2 \\ &\cong 35.10\text{ cm}^2 \end{aligned}$$

Substitute these two values into the rearranged modified one-group critical equation.

$$\begin{aligned} k_\infty &= 1 + B^2 M_T^2 \\ &= 1 + \left(1.54\times10^{-2}\text{ cm}^{-2}\right)\left(35.10\text{ cm}^2\right) \\ &= 1.54 \end{aligned}$$

Use this and the value for η_{TH} from Table 4.3 in App. 4.A.

$$\begin{aligned} k_\infty &= \eta_{\text{TH}} f \\ f &= \frac{k_\infty}{\eta_{\text{TH}}} \\ &= \frac{1.54}{2.065} \\ &= 0.746 \end{aligned}$$

The thermal utilization, which contains the mass term desired, is defined as follows.

$$\begin{aligned} f &= \frac{\overline{\Sigma}_{a,\text{fuel}}}{\overline{\Sigma}_{a,\text{mix}}} = \frac{\overline{\Sigma}_{a,\text{fuel}}}{\overline{\Sigma}_{a,\text{fuel}} + \overline{\Sigma}_{a,\text{H2O}}} \\ &= \frac{N_{\text{fuel}}\overline{\sigma}_{a,\text{fuel}}}{N_{\text{fuel}}\overline{\sigma}_{a,\text{fuel}} + N_{\text{H2O}}\overline{\sigma}_{a,\text{H2O}}} \\ &= \frac{\left(\frac{m_{\text{fuel}}N_A}{V\,MW_{\text{fuel}}}\right)\left(\frac{\sqrt{\pi}}{2}\right)\left(g_{\text{fuel}}(t)\right)\left(\sigma_{a,\text{fuel}}(E_0)\right)\left(\sqrt{\frac{T_0}{T}}\right)}{\left(\frac{m_{\text{fuel}}N_A}{V\,MW_{\text{fuel}}}\right)\left(\frac{\sqrt{\pi}}{2}\right)\left(g_{\text{fuel}}(t)\right)\left(\sigma_{a,\text{fuel}}(E_0)\right)\left(\sqrt{\frac{T_0}{T}}\right) + \left(\frac{m_{\text{H2O}}N_A}{V\,MW_{\text{H2O}}}\right)\left(\frac{\sqrt{\pi}}{2}\right)\left(g_{\text{H2O}}(t)\right)\left(\sigma_{a,\text{H2O}}(E_0)\right)\left(\sqrt{\frac{T_0}{T}}\right)} \end{aligned}$$

Since $T_0 = T = 20°\text{C}$, the ratio $T_0/T = 1$. The non-$1/v$ factor for the moderator is given as one. The volume V, Avogadro's number N_A, and the constant can be factored out and canceled. The result is

$$f = \frac{\left(\frac{m_{\text{fuel}}}{MW_{\text{fuel}}}\right)\left(g_{\text{fuel}}(T_0)\right)\left(\sigma_{a,\text{fuel}}(E_0)\right)}{\left(\frac{m_{\text{fuel}}}{MW_{\text{fuel}}}\right)\left(g_{\text{fuel}}(T_0)\right)\left(\sigma_{a,\text{fuel}}(E_0)\right) + \left(\frac{m_{\text{H2O}}}{MW_{\text{H2O}}}\right)\left(g_a(T_0)\right)\left(\sigma_{a,\text{H2O}}(E_0)\right)}$$

The non-1/v factor for the fuel g_{fuel} is obtained from Table 4.3 in App. 4.A as is the microscopic fission cross section for U-235 and the microscopic absorption cross section for the water moderator. Substituting gives

$$f = \frac{\left(\frac{m_{\text{fuel}}}{MW_{\text{fuel}}}\right)\left(g_{\text{fuel}}(T_0)\right)\left(\sigma_{a,\text{fuel}}(E_0)\right)}{\left(\frac{m_{\text{fuel}}}{MW_{\text{fuel}}}\right)\left(g_{\text{fuel}}(T_0)\right)\left(\sigma_{a,\text{fuel}}(E_0)\right) + \left(\frac{m_{\text{H2O}}}{MW_{\text{H2O}}}\right)\left(g_a(T_0)\right)\left(\sigma_{a,\text{H2O}}(E_0)\right)}$$

$$= \frac{\left(\frac{m_{\text{fuel}}}{235\,\frac{\text{g}}{\text{mol}}}\right)(0.9780)(681\text{ barns})}{\left(\frac{m_{\text{fuel}}}{235\,\frac{\text{g}}{\text{mol}}}\right)(0.9780)(681\text{ barns}) + \left(\frac{m_{\text{H2O}}}{18\,\frac{\text{g}}{\text{mol}}}\right)(1)(0.664\text{ barns})}$$

$$= \frac{(m_{\text{fuel}})(2.83\text{ barns})}{(m_{\text{fuel}})(2.83\text{ barns}) + (m_{\text{H2O}})(0.037\text{ barns})}$$

The units of barns and moles in the numerator and denominator will cancel. Substitute the calculated value for the thermal utilization and solve for the mass of the fuel.

$$f = \frac{2.83m_{\text{fuel}}}{2.83m_{\text{fuel}} + 0.037m_{\text{H2O}}}$$

$$0.746 = \frac{2.83m_{\text{fuel}}}{2.83m_{\text{fuel}} + 0.037m_{\text{H2O}}}$$

$$(0.746)(2.83m_{\text{fuel}} + 0.037m_{\text{H2O}}) = 2.83m_{\text{fuel}}$$

$$2.11m_{\text{fuel}} + 0.0276m_{\text{H2O}} = 2.83m_{\text{fuel}}$$

$$0.0276m_{\text{H2O}} = 0.72m_{\text{fuel}}$$

The final result is then

$$m_{\text{fuel}} = 3.83\times10^{-2}\,m_{\text{H2O}}$$

As in most homogeneous mixtures, the mass of the fissile material will be small. Therefore, the mass of the water can be calculated using its ordinary density.

Thus,

$$\rho_{\text{H2O}} = \frac{m_{\text{H2O}}}{V}$$

$$m_{\text{H2O}} = \rho_{\text{H2O}}V = \rho_{\text{H2O}}\pi R^2 H$$

$$= \left(\frac{1\text{ g}}{1\text{ cm}^3}\right)(\pi)(0.2\text{ m})^2(1\text{ m})\left(\frac{10^2\text{ cm}}{\text{m}}\right)^3$$

$$= 1.26\times10^5\text{ g}$$

Substituting gives

$$m_{\text{fuel}} = 3.83\times10^{-2}\,m_{\text{H2O}}$$

$$= (3.83\times10^{-2})(1.26\times10^5\text{ g})$$

$$= \boxed{4826\text{ g}\quad(4800\text{ g})}$$

Answer is D.

NUCLEAR THEORY–15

What is the flux at the center of the reactor (i.e., where the radius r = 0) at the core mid-plane (i.e., where z = 0) if the output power is 100 kW?

(A) 1.8×10^{12} cm^{-2}·s^{-1}
(B) 5.6×10^{12} cm^{-2}·s^{-1}
(C) 13.0×10^{12} cm^{-2}·s^{-1}
(D) 13.6×10^{12} cm^{-2}·s^{-1}

Solution:

The flux is given by the following.

$$\phi_{\text{TH}} = \text{A}J_0\left(\frac{2.405r}{R}\right)\cos\left(\frac{\pi z}{H}\right)$$

The constant A is given by

$$\text{A} = \frac{3.63P}{VE_R\Sigma_f}$$

Solve for the unknown items in the constant A, starting with the volume.

$$V = \pi R^2 H = \pi(0.2\text{ m})^2(1\text{ m}) = 0.126\text{ m}^3$$

$$\bar{\Sigma}_f = N_f \bar{\sigma}_f = \left(\frac{m_f N_A}{V\, MW}\right)\left(\frac{\sqrt{\pi}}{2}\right)(g_f(T))(\sigma_f(T_0))\left(\sqrt{\frac{T_0}{T}}\right) = \left(\frac{m_f N_A}{V\, MW}\right)\left(\frac{\sqrt{\pi}}{2}\right)(g_f(T_0))(\sigma_f(T_0))(1)$$

Substitute the values calculated in Prob. 14 and from Tables 4.1 and 4.3 in App. 4.A.

$$\bar{\Sigma}_f = \left(\frac{m_f N_A}{V\, MW}\right)\left(\frac{\sqrt{\pi}}{2}\right)(g_f(T_0))(\sigma_f(T_0))(1)$$

$$= \left(\frac{(4826\text{ g})\left(0.602\times 10^{24}\ \frac{\text{atoms}}{\text{mol}}\right)}{(0.126\text{ m}^3)\left(235\ \frac{\text{g}}{\text{mol}}\right)}\right)\left(\frac{\sqrt{\pi}}{2}\right)(0.9759) \times (585\text{ barns})\left(\frac{10^{-24}\text{ cm}^2}{1\text{ barn}}\right)\left(\frac{1\text{ m}}{10^2\text{ cm}}\right)^3$$

$$= 0.0496\text{ cm}^{-1}$$

Use the calculated and given values to determine the constant A.

$$\text{A} = \frac{3.63P}{VE_R\bar{\Sigma}_f}$$

$$= \frac{(3.63)(100\times 10^3\text{ W})\left(\frac{1\frac{\text{J}}{\text{s}}}{1\text{ W}}\right)}{(0.126\text{ m}^3)\left(\frac{10^2\text{ cm}}{1\text{ m}}\right)^3(200\text{ MeV})\left(\frac{10^6\text{ eV}}{1\text{ MeV}}\right)\times\left(\frac{1.602\times 10^{-19}\text{ J}}{1\text{ eV}}\right)(0.0496\text{ cm}^{-1})}$$

$$= 1.8\times 10^{12}\text{ neutrons/cm}^2\cdot\text{s}$$

All the information required to determine the flux is now available. Substitute this information into the flux equation. Obtain the value for the Bessel function from Table 3.1 of App. 3A.

$$\phi_{\text{TH}}(r) = \text{A}J_0\left(\frac{2.405r}{R}\right)\cos\left(\frac{\pi z}{H}\right)$$

$$\phi_{\text{TH}}(1\text{ m}) = \left(1.8\times 10^{12}\ \frac{\text{neutrons}}{\text{cm}^2\cdot\text{s}}\right)J_0\left(\frac{(2.405)(0\text{ m})}{0.2\text{ m}}\right)\times\cos\left(\frac{\pi(0\text{ m})}{1\text{ m}}\right)$$

$$= \left(1.8\times 10^{12}\ \frac{\text{neutrons}}{\text{cm}^2\cdot\text{s}}\right)J_0(0)\cos(0)$$

$$= \boxed{1.8\times 10^{12}\text{ neutrons/cm}^2\cdot\text{s}}$$

Answer is A.

NUCLEAR THEORY–16

A reflector has what effect(s)?

(A) flattens the thermal flux
(B) flattens the fast flux
(C) reduces the critical mass
(D) A and C

Solution:

A reflector is a layer of scattering material (i.e., unfueled moderator) around a core which is designed to lower the loss of neutrons.

In graphic form the effects are as follows.

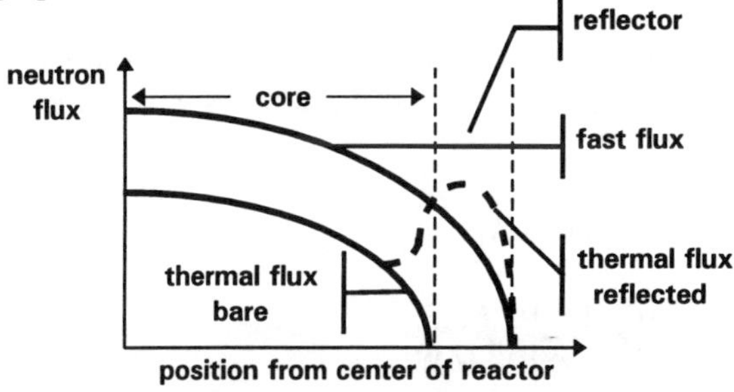

As can be seen from the diagram, the effect on the thermal flux is one of flattening. A flatter thermal flux improves the performance of the core by providing for a more even fuel burnup and allowing operation further from thermal limits, among other advantages.[24] Since some of the neutrons that would normally leak from the core instead diffuse back into the core from the reflector, less fuel is required for a critical mass. In terms of equations, the buckling B^2, which represents the curvature of the flux, is less (i.e., flatter); therefore, the non-leakage probability is larger.

$$P_{NL}\Uparrow = \frac{1}{1+B^2 M_T^2\downarrow}$$

The multiplication factor can then be smaller and maintain criticality.

$$\Downarrow k_\infty P_{NL}\downarrow = 1$$

This means the thermal utilization f can be less.

$$k_\infty\downarrow = \eta_{\text{TH}} f \Downarrow \varepsilon\phi$$

Which means less fuel is required.

$$f\downarrow = \frac{\Sigma_{a,\text{fuel}}\Downarrow}{\Sigma_{a,\text{mix}}}$$

Answer is D.

NUCLEAR THEORY–17

The reactor period T for an infinite thermal reactor, without delayed neutrons, is given by

$$T = \frac{\ell_p}{k_\infty - 1}$$

ℓ_p = prompt neutron lifetime[25]

If the reactivity ρ changes by 0.1% in a critical thermal reactor operating at 100 MW, what is the power one second later?

(A) 110 MW
(B) 400 MW
(C) 2×10^6 MW
(D) 3×10^{45} MW

Solution:

The reactivity is defined as follows.[26]

$$\rho = \frac{k_\infty - 1}{k_\infty}$$

A change of 0.1% is equivalent to a change of 0.001. Therefore,

$$\begin{aligned}\rho &= \frac{k_\infty - 1}{k_\infty}\\ k_\infty\rho &= k_\infty - 1\\ k_\infty(\rho - 1) &= -1\\ k_\infty &= \frac{-1}{\rho - 1}\\ &= \frac{-1}{0.001 - 1}\\ &= 1.001\end{aligned}$$

Solving for the reactor period gives

$$\begin{aligned}T &= \frac{\ell_p}{k_\infty - 1}\\ &= \frac{10^{-4}\ \text{s}}{1.001 - 1}\\ &= 0.1\ \text{s}\end{aligned}$$

The reactor power can now be determined using the following formula.

$$P_f = P_0 e^{\frac{t}{T}}$$

P_f = final power
P_0 = initial power
t = elapsed time

Solve for the final power.

[24] The maximum to average flux ratio is reduced and the critical dimensions are reduced. The dimension decrease is called the *reflector savings* and is defined based on the geometry.
[25] The value for the prompt lifetime is approximately 10^{-4} s in water moderated thermal reactors and 10^{-7} s in fast reactors.
[26] This formula takes the same form for finite reactors.

$$P_f = P_0 e^{\frac{t}{T}}$$

$$= (100\text{ MW})\exp\left(\frac{1\text{ s}}{0.1\text{ s}}\right)$$

$$= \boxed{2.2\times10^6\text{ MW} \quad (2\times10^6\text{ MW})}$$

Answer is C .

Note: The important principle here is that without the delayed neutrons, the reactor would be uncontrollable. With delayed neutrons involved the reactor period is on the order of one minute for the same reactivity change.[27]

NUCLEAR THEORY–18

The reactivity in a reactor fueled with U-235 changes by 0.2% due to an outward rod shim. What is the reactivity change in cents?

(A) 2 cents
(B) 31 cents
(C) 77 cents
(D) 102 cents

Solution:

The 0.2% change represents a reactivity change of 0.002. A *dollar* of reactivity is defined as the amount of reactivity necessary to take a reactor *prompt critical* (i.e., critical on prompt neutrons alone). In equation form this means

$$\rho = \beta$$

β = delayed neutron fraction; the fraction of all fission neutrons born delayed

From Table 4.5 in App. 4.B, the value of β for a U-235 fueled reactor is 0.0065. Therefore, one dollar of reactivity is equal to 0.0065. The given reactivity in dollars is then

$$\rho = \frac{0.002}{0.0065} = 0.308\text{ dollars}$$

A cent of reactivity is 1/100 of a dollar. Thus,

$$\rho = (0.308\text{ dollars})\left(\frac{100\text{ cents}}{1\text{ dollar}}\right)$$

$$= \boxed{30.8\text{ cents} \quad (31\text{ cents})}$$

Answer is B.

Note: Since a dollar of reactivity is defined in terms of the delayed neutron fraction, which varies depending on the fuel, it is not a constant value.

NUCLEAR THEORY–19

Consider the following table of differential rod worth and integral rod worth.

DRW $\Delta\rho/\rho$	IRW $\rho(x)/\rho(H)$ \$	rod position (H) (arbitrary units)
0	0	0
7.48×10^{-3}	1.15	25
4.00×10^{-3}	6.15	50
7.35×10^{-2}	11.30	75
8.00×10^{-2}	12.30	100

DRW = differential rod worth
IRW = integral rod worth

The delayed neutron fraction β is taken to be 0.0065. It takes three minutes for a rod to move from the bottom to the top of the core. Safety considerations and regulations limit the reactivity insertion rate to $14\times10^{-4}\ \Delta\rho/\rho$ or approximately 21 cents per second during steady state conditions.

Consider the following curves of rod worth.

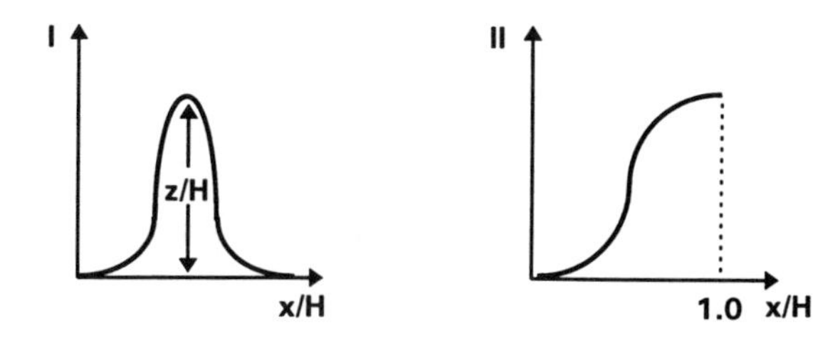

[27] This is based on using six delayed neutron precursor groups.

What is the maximum reactivity insertion rate? Which curve shown represents the shape of a differential rod worth curve?

(A) 3 cents/s I
(B) 9 cents/s II
(C) 12 cents/s I
(D) 42 cents/s II

Solution:

First, the differential rod worth curve (DRW) is curve I. The rod worth changes very little at either the top or bottom of the core primarily due to the low neutron flux in these regions. By contrast, it changes most near the center of the core due to the high neutron flux present in this region.

Curve II starts with a value of zero and builds. This can only be the integral rod worth (IRW) curve.

The maximum reactivity change takes place between the 50 and 75 unit positions ($5.15). The reactivity change in this region can be confirmed as follows.

$$\begin{aligned}\Delta\rho &= \rho_{75} - \rho_{50} \\ &= 7.35\times 10^{-2} - 4.00\times 10^{-2} \\ &= 3.35\times 10^{-2}\end{aligned}$$

In terms of dollars,

$$\begin{aligned}\Delta\rho &= \frac{3.35\times 10^{-2}}{0.0065} \\ &= \$5.15\end{aligned}$$

Solve for the reactivity change per height unit.

$$\begin{aligned}\frac{\Delta\rho}{\Delta H} &= \frac{\$5.15}{75\text{ units} - 50\text{ units}} \\ &= 0.206\frac{\$}{\text{height unit}} \\ &= 21\text{ cents/height unit}\end{aligned}$$

The rod speed (RS) is calculated from the following.

$$\begin{aligned}RS &= \frac{\Delta\text{ unit height}}{\text{time}} \\ &= \frac{100\text{ height units}}{(3\text{ min})\left(\frac{60\text{ s}}{1\text{ min}}\right)} \\ &= 0.556\text{ height units/s}\end{aligned}$$

Therefore, the reactivity change per second (i.e., the maximum reactivity insertion rate) is as follows.

$$\begin{aligned}\left.\frac{\Delta\rho}{\Delta t}\right|_{\max} &= \left(\left.\frac{\Delta\rho}{\Delta H}\right|_{\max}\right)(RS) \\ &= \left(21\frac{\text{cents}}{\text{height unit}}\right)\left(0.556\frac{\text{height units}}{\text{s}}\right) \\ &= \boxed{11.7\text{ cents/s}\quad (12\text{ cents/s})}\end{aligned}$$

Answer is C.

NUCLEAR THEORY–20

The concentration of Xe-135, a fission product poison, is given by the following.

$$\frac{dN_{\text{Xe}}}{dt} = \lambda_I N_I + \gamma_{\text{Xe}}\overline{\Sigma}_f\phi_{\text{TH}} - \lambda_{\text{Xe}}N_{\text{Xe}} - \overline{\Sigma}_{a,\text{Xe}}\phi_{\text{TH}}$$

N_{se} = atom density of xenon
λ_I = decay constant of iodine
N_I = atom density of iodine
γ_{Xe} = fission yield of xenon
Σ = macroscopic cross section
ϕ_{TH} = thermal flux

Consider the following graph of xenon concentration over time.

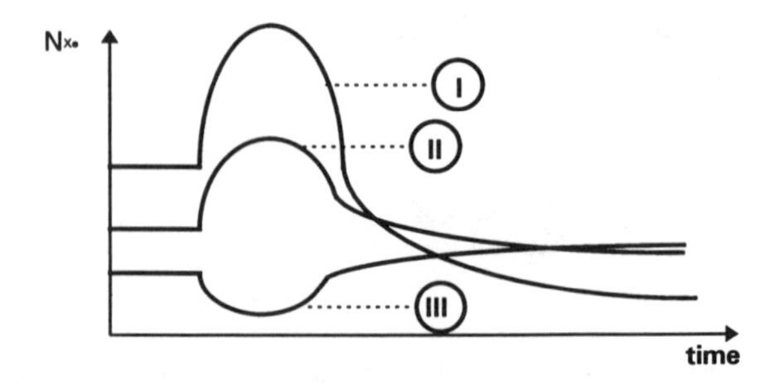

Which curve represents an up-power, down-power, and shutdown transient?

	up	down	shutdown
(A)	I	II	III
(B)	II	III	I
(C)	III	I	II
(D)	III	II	I

Solution:

When a reactor is shutdown, the production of xenon directly from fission ends.

$$\gamma_{\text{Xe}}\overline{\Sigma}_f\phi_{\text{th}} = 0$$

The absorption of neutrons by xenon also ends as the thermal flux is assumed to be zero.

$$\overline{\Sigma}_{a,\text{Xe}}\phi_{\text{th}} = 0$$

Therefore, xenon is being produced by the decay of iodine and concentration of xenon increases.

$$\lambda_I N_I$$

Once produced, the xenon begins to decay.

$$\lambda_{\text{Xe}} N_{\text{Xe}}$$

Additionally, the decay constant for iodine is larger than the decay constant of xenon (i.e., $2.93 \times 10^{-5}\ \text{s}^{-1}$ versus $2.11 \times 10^{-5}\ \text{s}^{-1}$).

The net result is an initial increase in the concentration of xenon with a minor loss, due to the decay of xenon, taking place gradually.

Thus, curve I represents the shutdown transient.

During a down-power transient the magnitude of the thermal flux ϕ_{th} decreases. This causes a decrease in the minor production factor term.

$$\gamma_{\text{Xe}}\overline{\Sigma}_f\phi_{\text{th}}$$

This term is minor considering the thermal fission yield for xenon is 0.00237 while for iodine it is 0.0639, about thirty times smaller.

This decrease in the thermal flux also results in a decrease in the major loss term.

$$\overline{\Sigma}_{a,\text{Xe}}\phi_{\text{th}}$$

The net result is an initial increase in the xenon concentration until equilibrium is again reached.

Thus, curve II represents a down power transient.

By similar reasoning an up-power transient increases the minor production term and the major loss term resulting in an initial xenon concentration decrease.

Thus curve III represents an up-power transient.

Answer is D.

11 NUCLEAR INSTRUMENTS: SOLUTIONS

NUCLEAR INSTRUMENTS–1

The general method of producing x-rays from target material for use in radiography or for medical purposes is shown.

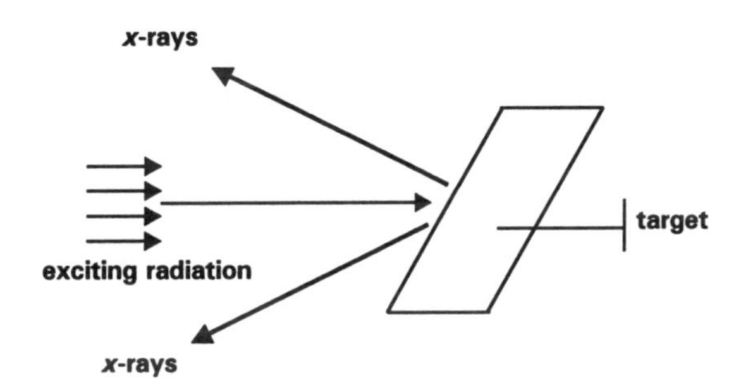

The exciting radiation can be an ionizing radiation.[1] Assume excitation of the target results in the expulsion of a K-shell electron which is then filled by an L-shell electron as shown.[2]

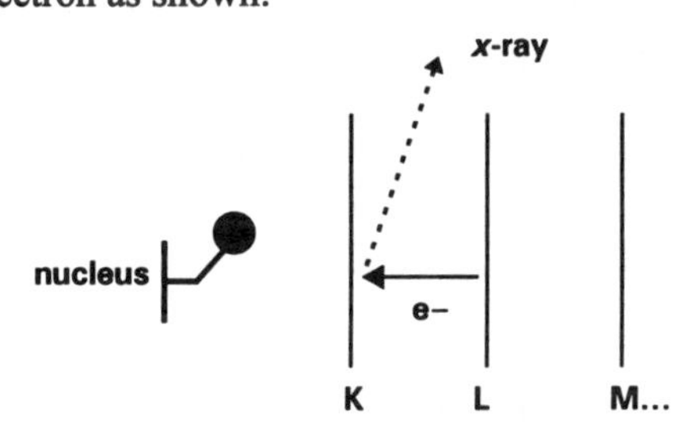

The emitted x-ray is called by what term?

(A) Auger x-ray
(B) internal conversion x-ray
(C) Bremsstrahlung x-ray
(D) K_α characteristic x-ray

[1] The range under discussion is from approximately 10 eV, the minimum energy required to cause ionization in typically used materials, to about 20 MeV which is the upper bound of energy for concern in nuclear science and technology.
[2] The labels K, L, M, and so on represent the total quantum numbers 1, 2, 3, and so on. The designations s, f, p, d, and so on are representative of angular-momentum states.

Solution:

The term Auger refers to an electron emitted when an atom interacts with its outer shell electron to transfer excitation energy. This results in a free electron. The process is analogous to internal conversion, the difference being that internal conversion is an interaction with the nucleus while an Auger electron results from an interaction with the atom as a whole. The electron emitted possesses the difference between the excitation energy of the atom and the binding energy of the electron. The energy spectrum of Auger electrons is thus discrete and can be useful in radiation instruments. Therefore, A is not the answer.

When a nucleus is in an excited state and transfers its energy to one of its electrons, thereby eliminating it, this is called internal conversion. They are a useful high keV to MeV electron source, though the spectrum of energy is not discrete. Thus, B is not the answer.

Bremsstrahlung radiation is of a x-ray nature but emanates from fast electrons slowing and shedding excess energy by emitting. The energy spectrum is a continuum and varies depending upon the material in which the electron is slowed. Therefore, C is not the answer.

When an outer electron drops into a lower energy level inner shell, it emits its energy as a characteristic x-ray. The x-ray is characteristic in that it is discrete for a particular shell to shell transition and for the element involved, as only certain transitions are possible depending upon the electron shell configuration. Transition from the L to K shell are termed K_α, from M to K are K_β, and so on.

Answer is D.

NUCLEAR INSTRUMENTS–2

Neutron sources are required for the testing of nuclear instrumentation and for the calibration of neutron radiation detectors. Two significant radioisotope photoneutron sources follow.

$$^{9}_{4}\text{Be} + h\nu \rightarrow ^{8}_{4}\text{Be} + ^{1}_{0}\text{n}$$

$$^{2}_{1}\text{H} + h\nu \rightarrow ^{1}_{1}\text{H} + ^{1}_{0}\text{n}$$

The Q value for the beryllium reaction is –1.666 MeV. The Q value for the hydrogen reaction is –2.226 MeV.

If the beryllium source is used, what type of radiation, based on the frequency required, is necessary to cause the reaction to occur?

(A) radio
(B) infrared
(C) visible
(D) gamma

Solution:

The Q value is negative and thus the process is endothermic.[3] That is, energy must be added to make the reaction occur. The energy comes from the incident particle and is given by the following formula.

$$E = h\nu$$

h = Planck's constant
ν = frequency

Substitute the given information and the value for Planck's constant.

$$\begin{aligned} E &= h\nu \\ \nu &= \frac{E}{h} \\ &= \frac{(1.666\ \text{MeV})\left(\frac{10^{6}\ \text{eV}}{1\ \text{MeV}}\right)\left(\frac{1.602\times 10^{-19}\ \text{J}}{1\ \text{eV}}\right)}{6.6256\times 10^{-34}\ \text{J}\cdot\text{s}} \\ &= 4.03\times 10^{20}\ \text{s}^{-1} \end{aligned}$$

In terms of the wavelength, for a particle of zero rest mass, the applicable calculation is

$$\begin{aligned} \lambda &= \frac{hc}{E} \\ &= \frac{(6.6256\times 10^{-34}\ \text{J}\cdot\text{s})\left(3\times 10^{8}\ \frac{\text{m}}{\text{s}}\right)}{(1.66\ \text{MeV})\left(1.602\times 10^{-13}\ \frac{\text{J}}{\text{MeV}}\right)} \\ &= (7.47\times 10^{-13}\ \text{m})\left(\frac{1\ \text{Å}}{10^{-10}\ \text{m}}\right) \\ &= 7.47\times 10^{-3}\ \text{Å} \end{aligned}$$

Note: The angstrom is a common unit for wavelength.

Using Table 5.1 of App. 5A shows this frequency and wavelength to be in the range of an x-ray or γ-ray. The source was not specified thus the only available answer is gamma.[4]

Answer is D.

NUCLEAR INSTRUMENTS–3

A source and background count of 760 counts over a three minute period is measured by a laboratory detector. A stable average background count taken over a long period of time measures 100 counts per minute (cpm).

What is the net count rate and standard deviation for the source alone?

(A) 150 cpm ± 9 cpm
(B) 660 cpm ± 9 cpm
(C) 150 cpm ± 13 cpm
(D) 660 cpm ± 13 cpm

Solution:

The net count rate (ncr) is

$$\text{net count rate} = \frac{\text{total counts}}{t} - \frac{\text{background counts}}{t}$$

[3] This Q value is also called the decay energy, reaction energy, and sometimes the threshold energy.

[4] Recall that gamma rays are generated in the nucleus. The generating source of x-rays is orbital electrons.

Substitute the given numbers.

$$ncr = \frac{760\ \text{counts}}{3\ \text{min}} - 100\ \text{cpm}$$
$$= 153.3\ \text{cpm} \quad (150\ \text{cpm})$$

The answer is A or C.

The standard deviation for the source and background count is calculated as follows.

$$\sigma_{s+bg} = \frac{\sqrt{\text{counts}}}{t}$$
$$= \frac{\sqrt{760\ \text{counts}}}{3\ \text{min}}$$
$$= 9.2\ \text{min}^{-1} \quad (9\ \text{min}^{-1})$$

Since the background count is stable and taken over a long period of time, no deviation is calculated.[5] The final result is

$$\boxed{150\ \text{cpm} \pm 9\ \text{cpm}}$$

Answer is A.

NUCLEAR INSTRUMENTS–4

A 100 cm^2 swipe is taken from an area suspected of harboring surface contamination. The result of an initial one minute count is

125 counts

Due to the unexpectedly high results, two additional counts are completed of three minutes duration. The results are as follows.

350 counts
362 counts

The one minute background count completed just prior to the survey indicated 20 counts.

A report is to be completed based on this information. What is the net average count rate per minute and the standard deviation in counts per minute (cpm)?

(A) 100 cpm ± 13 cpm
(B) 120 cpm ± 12 cpm
(C) 120 cpm ± 13 cpm
(D) 360 cpm ± 13 cpm

Solution:

First, the total average count rate is

$$\frac{\text{total cpm}}{\text{number of counts}} = \frac{\text{count rate}_1 + \text{count rate}_2 + \text{count rate}_3}{3}$$
$$= \frac{\left(\frac{125\ \text{counts}}{1\ \text{min}}\right) + \left(\frac{350\ \text{counts}}{3\ \text{min}}\right) + \left(\frac{362\ \text{counts}}{3\ \text{min}}\right)}{3}$$
$$= \frac{125\ \text{cpm} + 116.7\ \text{cpm} + 120.7\ \text{cpm}}{3}$$
$$= 120.8\ \text{cpm}$$

From this result the background count must be subtracted to obtain the net count rate. The background count rate is

$$\text{background cpm} = \frac{\text{background count}}{t}$$
$$= \frac{20\ \text{counts}}{1\ \text{min}}$$
$$= 20\ \text{cpm}$$

The net count rate may now be determined.

$$\text{net cpm} = \text{total cpm} - \text{background cpm}$$
$$= 120.8\ \text{cpm} - 20\ \text{cpm}$$
$$= 100.8\ \text{cpm} \quad (100\ \text{cpm})$$

The *standard deviation* σ is the square root of the *variance* σ^2.

[5] The time frame of infinity would have to occur to make this strictly true mathematically.

For the initial count,

$$\sigma_1 = \frac{\sqrt{\text{counts}}}{t} = \frac{\sqrt{125\ \text{counts}}}{1\ \text{min}} = 11.18\ \text{min}^{-1}$$

The second and third counts may be treated individually, as is always the case, or as one count since the counting time is identical.[6] Using the latter approach gives

$$\sigma_{2,3} = \frac{\sqrt{\text{total counts}}}{\text{total } t} = \frac{\sqrt{350\ \text{counts} + 362\ \text{counts}}}{3\ \text{min} + 3\ \text{min}} = 4.45\ \text{min}^{-1}$$

Since the count rates 1, 2, and 3 were added to get the total count rate, the standard deviations must be combined as shown.

$$\sigma_{\text{tot}} = \sqrt{\sigma_1{}^2 + \sigma_{2,3}{}^2} = \sqrt{\left(11.18\ \text{min}^{-1}\right)^2 + \left(4.45\ \text{min}^{-1}\right)^2} = 12.03\ \text{min}^{-1}$$

The standard deviation for the background count is calculated as follows.

$$\sigma_{bg} = \frac{\sqrt{\text{background counts}}}{t} = \frac{\sqrt{20\ \text{counts}}}{1\ \text{min}} = 4.47\ \text{min}^{-1}$$

Since the background count rate was subtracted from the total count rate to obtain the net count rate, the standard deviation for the net count rate is determined as shown.[7]

$$\sigma_{\text{net}} = \sqrt{\sigma_{\text{tot}}{}^2 + \sigma_{bg}{}^2} = \sqrt{\left(12.03\ \text{min}^{-1}\right)^2 + \left(4.47\ \text{min}^{-1}\right)^2} = 12.8\ \text{min}^{-1} \quad (13\ \text{cpm})$$

The overall result is

100 cpm ± 13 cpm

Answer is A.

Note: This entire process is called *error propagation.* Additionally, a gaussian or normal distribution was assumed.

NUCLEAR INSTRUMENTS–5

Three statistical models used in nuclear engineering are the binomial, poisson, and gaussian distributions. The first is used in constant probability processes; the second where the probability of success is small (e.g., long half-life, low observation time); the third where the probability of success is large (e.g., 20 or more counts).

If the definition of success is the decay of a nucleus during an observation, what is the probability of success during a five minute observation of a radioactive source with a half-life of 53 minutes?

(A) 0.06
(B) 0.09
(C) 0.91
(D) 0.94

Solution:

Recall from the Nuclear Radiation section the following formula.

$$n(t) = n_0 e^{-\lambda t}$$

The factor $\exp(-\lambda t)$ represents the fraction of original atoms that have not decayed. Therefore, the fraction of those that have decayed is

$$1 - e^{-\lambda t}$$

This is also the definition of the probability of success (p).

[6] It is important to realize that this works only if the counting time frames are equal. If not, handle them as separate counts as will be shown in the next step of this problem.

[7] Even though subtraction occurs, the standard deviations are added in the manner shown. This stems from the squares in the error propagation formula used. See the Reference section on radiation detection for additional information.

Therefore, the desired quantity can be determined from the following.

$$
\begin{aligned}
p &= 1 - e^{-\lambda t} \\
&= 1 - \exp\left(\left(\frac{-0.693}{T_{1/2}}\right)(t)\right) \\
&= 1 - \exp\left(\left(\frac{-0.693}{53\,\text{min}}\right)(5\,\text{min})\right) \\
&= \boxed{0.06}
\end{aligned}
$$

Answer is A.

NUCLEAR INSTRUMENTS–6

A simplified detector and output circuitry is shown.

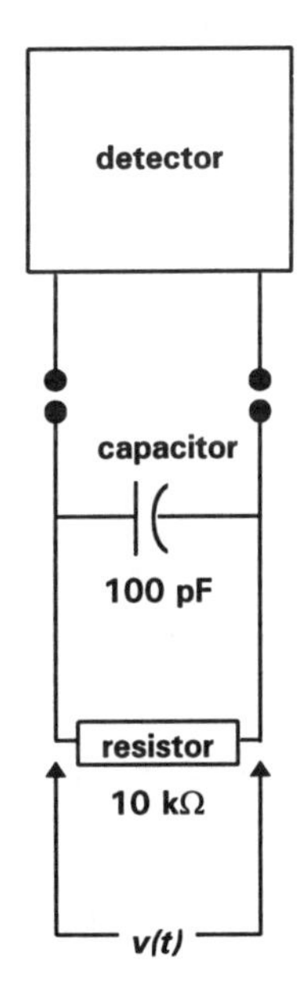

If the time constant λ is much larger than the detector charge collection time (i.e., a large value of RC), the output signal for a single interaction in the detector is as follows.

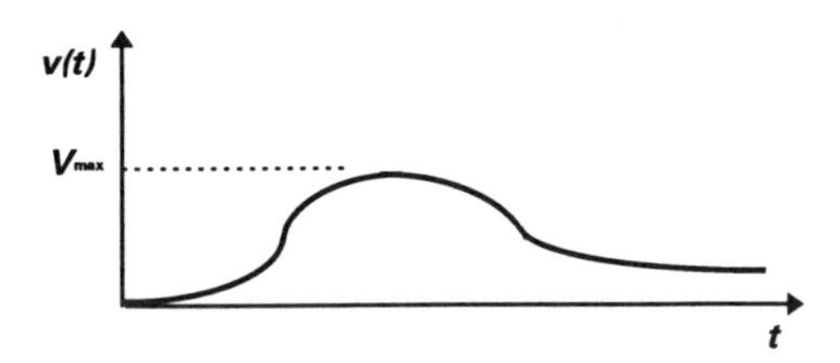

Assume one interaction in the detector creates 10^5 electrons of charge. What mode is the detector operating in and what is the approximate value of V_{max}?

(A) mean square voltage; 10 μV
(B) Campbelling; 0.02 V
(C) current; 10 μV
(D) pulse; 0.02 V

Solution:

The capacitance is defined as

$$C = \frac{\text{charge}}{\text{voltage}}$$

Capacitance is related to the current and voltage by

$$i = C\frac{dv}{dt}$$

Using the definition and limiting the value of concern to V_{max} thereby eliminating time concerns gives the following.[8]

$$
\begin{aligned}
C &= \frac{Q}{V_{max}} \\
V_{max} &= \frac{Q}{C}
\end{aligned}
$$

The value of the charge for a single interaction can be calculated as follows.

$$
\begin{aligned}
Q &= \left(\begin{array}{c}\text{number of electons} \\ \text{produced}\end{array}\right)\left(\frac{\text{charge}}{\text{electron}}\right) \\
&= \left(10^5\ \text{electrons}\right)\left(\frac{1.602\times10^{-19}\ \text{C}}{1\,\text{electron}}\right) \\
&= 1.602\times10^{-14}\ \text{C}
\end{aligned}
$$

[8] The long time constant allows the assumption that all the charge in the detector is collected by the capacitor and contributes to V_{max}. This is not strictly true for any RC circuit with a time constant less than infinity.

Substitute this value of the charge to determine V_{max}.

$$V_{max} = \frac{Q}{C}$$

$$= \frac{1.602\times10^{-14}\ \mathrm{C}}{\left(1\times10^{-12}\ \mathrm{F}\right)\left(\frac{1\frac{\mathrm{C}}{\mathrm{V}}}{1\ \mathrm{F}}\right)}$$

$$= \boxed{1.6\times10^{-2}\ \mathrm{V}\ \ (0.02\ \mathrm{V})}$$

Thus, the answer is either B or D.

As for the mode of operation, mean square voltage and Campbelling are identical. Instruments use this mode to differentiate between radiation types based on the charge they produce, since the electronic circuitry squares the resulting charge input. Squaring and averaging circuits are not shown. Thus, A and B are not the answer.

The current mode averages the fluctuations between individual radiation interactions to provide a relatively constant output. This mode is useful when the event rate (i.e., radiation interactions with the detector) is high; further, it's not constant with time. Therefore, C is not the answer.

The pulse mode is the most common mode of operation. It has the advantage of preserving information on the amplitude and timing of individual radiation events. The circuitry shown is of this type.

Answer is D.

NUCLEAR INSTRUMENTS–7

Iron-59 is believed present in a sample. To confirm, a radiation detector is set to measure the energy of the two gammas emitted during the decay expected. The gammas have an approximate energy of 1.30 MeV and 1.10 MeV.

What resolution is required by the detector to ensure these gamma energies can be detected as separate entities?

(A) 2%
(B) 10%
(C) 17%
(D) 65%

Solution:

Consider the following graph defining resolution.[9]

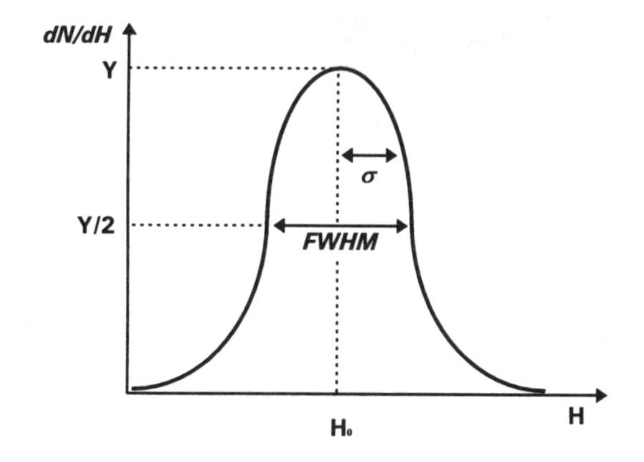

The resolution is defined from the following equation.

$$R = \frac{FWHM}{H_0}$$

$$FWHM = 2.38\sigma$$

N = number of pulses at a given pulse height
H = pulse height
H_0 = peak centroid
σ = standard deviation
$FWHM$ = full width at half maximum

In order to resolve two energy peaks the energies should be separated by at least one value of the *FWHM*.[10]

This means that the *FWHM* value should be the following.

$$FWHM = 1.30\ \mathrm{MeV} - 1.10\ \mathrm{MeV} = 0.20\ \mathrm{MeV}$$

The centroid H_0 is the value midway between these two values. Thus,

$$H_0 = 1.10\ \mathrm{MeV} + \frac{1.30\ \mathrm{MeV} - 1.10\ \mathrm{MeV}}{2} = 1.20\ \mathrm{MeV}$$

[9] This graph assumes a gaussian shape for the curve.
[10] This is an approximate rule of thumb.

Solve for the resolution with the calculated information.

$$R = \frac{FWHM}{H_0} = \frac{0.20\ \text{MeV}}{1.20\ \text{MeV}}$$

$$= \boxed{0.167 \text{ or } 16.7\% \quad (17\%)}$$

Note: This is not a high resolution. Semiconductor detectors' resolution can be less than 1%. Scintillation gamma ray detectors have resolutions between 5–10%. Proportional counters' resolution is roughly 10–15%.

Answer is C.

NUCLEAR INSTRUMENTS–8

A small DC ion chamber device known as pocket chamber or pocket dosimeter is often used to monitor exposure. They are fitted with an integral fiber electroscope which can be read when held up to a light. The initial charge zeroes the scale. The scale moves when exposed to a radiation field which results in a decrease of voltage in the chamber. A generic setup for such an instrument is shown.

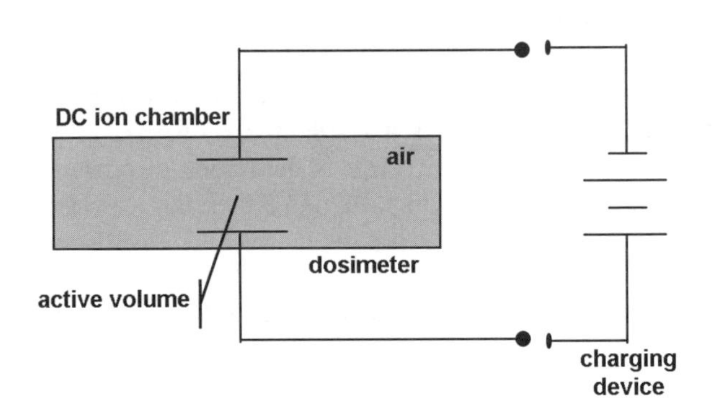

The active volume is 15 cm^3. The capacitance of the device is 75 pF. The battery voltage is 6 V. The air is at STP.[11]

What gamma ray exposure will reduce the initial chamber voltage to 4 V?

(A) 600 μR
(B) 30 mR
(C) 1 R
(D) 100 R

Solution:

Recall that the capacitance is given by

$$C = \frac{Q}{V}$$

Note that the unit farad equals coulombs/volt. Thus, the charge on the capacitor associated with the 6 V initial charge is[12]

$$\begin{aligned} C &= \frac{Q}{V} \\ Q_6 &= CV \\ &= (75 \times 10^{-12}\ \text{F})(6\ \text{V}) \\ &= \left(75 \times 10^{-12}\ \frac{\text{C}}{\text{V}}\right)(6\ \text{V}) \\ &= 4.5 \times 10^{-10}\ \text{C} \end{aligned}$$

The charge associated with 4 V is

$$\begin{aligned} Q_4 &= CV \\ &= (75 \times 10^{-12}\ \text{F})(4\ \text{V}) \\ &= 3.0 \times 10^{-10}\ \text{C} \end{aligned}$$

The difference represents the charge that must be produced in the chamber to lower the voltage by 2 V.[13]

$$\begin{aligned} Q_{\Delta 2} &= Q_6 - Q_4 \\ &= 4.5 \times 10^{-10}\ \text{C} - 3.0 \times 10^{-10}\ \text{C} \\ &= 1.5 \times 10^{-10}\ \text{C} \end{aligned}$$

Recall that one roentgen equals 2.58×10^{-4} C/kg.

[11] The definition of STP, standard temperature and pressure, actually varies. The one most encountered in physics is 0°C and one atmosphere (i.e., 760 torr). The density of one cubic centimeter of dry air is then 0.001293 g.

[12] The correct symbol for coulomb in the SI system is C. It is spelled in some instances here to avoid confusion with the capacitance.

[13] This could also have been solved for directly by using 2 V in the formula.

Thus, the exposure X necessary to change the voltage by 2 V can now be determined.[14]

$$X_{\Delta 2} = \frac{\left(1.5\times10^{-10}\ \text{C}\right)\left(\dfrac{1\ \text{R}}{2.58\times10^{-4}\ \dfrac{\text{C}}{\text{kg}}}\right)\left(\dfrac{10^3\ \text{g}}{1\ \text{kg}}\right)}{\left(15\ \text{cm}^3\right)\left(0.001293\ \dfrac{\text{g}}{\text{cm}^3}\right)}$$

$$= \boxed{0.03\ \text{R}\ \ (30\ \text{mR})}$$

Answer is B.

NUCLEAR INSTRUMENTS–9

The characteristics of gas-filled detectors operated in the pulse mode can represented by the following graph.

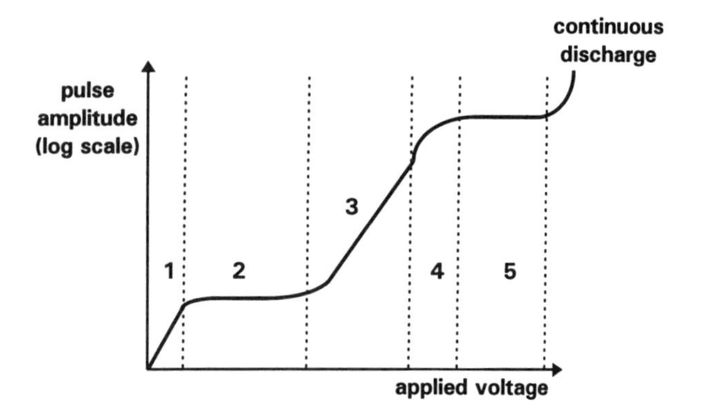

The regions under consideration are numbered 1–5. Consider now the following characteristics.

designation	characteristics
I	DC ion chamber region of operation
II	true proportionality; conventional proportional counters region of operation
III	recombination of created ion pairs occurs before collection
IV	space charge effects can come into play altering the electric field of the detector
V	a single ionizing event saturates the detector

Match the characteristic with the correct region.

(A) I 1 II 3 III 4 IV 5 V 2
(B) I 2 II 3 III 1 IV 4 V 5
(C) I 2 II 4 III 1 IV 3 V 5
(D) I 5 II 4 III 3 IV 2 V 1

Solution:

In region 1 the ion pairs formed during radiation interaction recombine prior to being collected. Region 1 therefore correlates with designation III.

In region 2 ion saturation occurs. The radiation interaction results in a given amount of charge which is fully collected. This is where DC ion chambers operate. Region 2 therefore correlates with designation I.

In region 3 gas multiplication occurs which is proportional to the original charge created by the radiation interaction. Region 3 therefore correlates with designation II.

In region 4 non-linear effects come into being, primarily space-charge effects. This effect occurs due to the slower moving positive ions not being collected at the same rate as their electron counterparts. The result is a distortion of the ion chamber electric field which eliminates the proportionality of region 3. Region 4 therefore correlates with designation IV.

In region 5 a single radiation interaction completely ionizes the chamber. As a result, the properties of the radiation are lost entirely. This is the region of operation of Geiger-Mueller detectors. Region 5 therefore correlates with designation V.

The correct combination is then I-2, II-3, III-1, IV-4, and V-5.

Answer is B.

[14] This formula was derived from unit analysis.

Nuclear Instruments (NI) problems 10–13 are based on the following information, drawing, and graph.

Crystalline sodium iodide with a trace of thallium iodide NaI[Tl] is the most common material for scintillation spectrometry.[15] Radiation interacts with the material exciting the individual atoms of the structure. They undergo scintillation producing a flash of light.[16] This light is coupled directly or via light pipes to a photomultiplier (PM) tube.[17] A photocathode sensing the weak scintillation light signal, produces free electrons which are then amplified in number in a multiplier section.

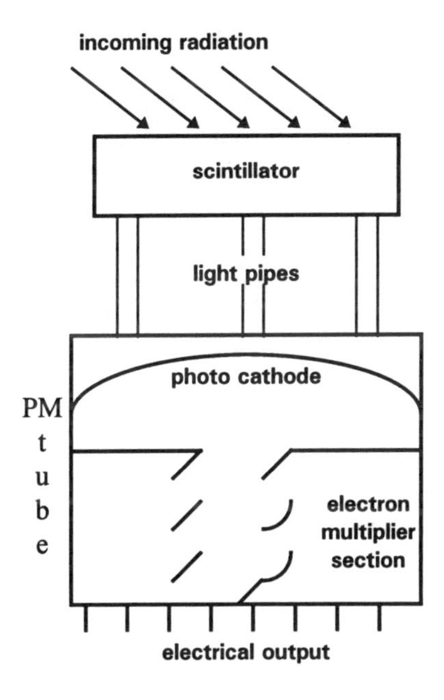

The electrical output, which goes to a multichannel analyzer (MCA), is shown on the next graph.[18] Numerous properties are included.[19] Different size detectors with multiple energy ranges are shown on the same graph.

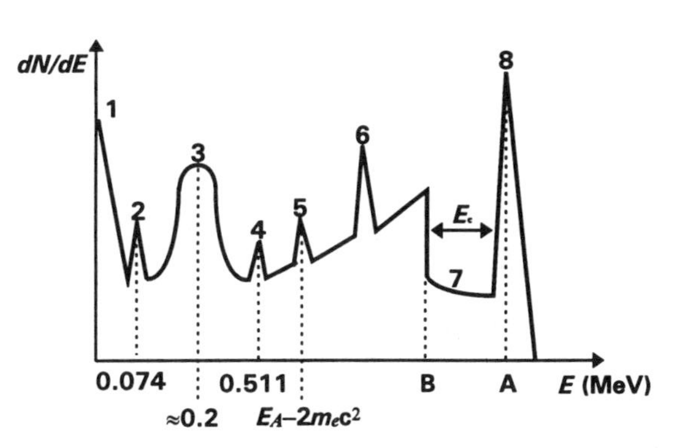

N = number of pulses
E = energy

NUCLEAR INSTRUMENTS–10

A given photocathode gain is 10^6 and the electron output pulse from the PM tube occurs in 6 ns.[20] What is the approximate peak pulse current output of the detector if incoming radiation causes a scintillation event releasing 1100 photoelectrons from the photocathode?

(A) 0.03 mA
(B) 0.30 mA
(C) 3.00 mA
(D) 30.0 mA

Solution:

Current has units of C/s or amperes A. The pulse charge and the peak current are found as follows.

$$Q = \begin{pmatrix} \text{number of} \\ \text{photoelectrons} \end{pmatrix} (\text{gain}) \left(\frac{\text{charge}}{\text{electron}} \right)$$

$$= (1100\,\text{electrons})(10^6)\left(1.602 \times 10^{-19}\ \frac{\text{C}}{\text{electron}}\right)$$

$$= 1.76 \times 10^{-10}\ \text{C}$$

$$I_{\text{peak}} = \frac{Q}{t}$$

$$= \frac{1.76 \times 10^{-10}\ \text{C}}{6 \times 10^{-9}\ \text{ns}}$$

$$= \boxed{0.029\ \text{A}\quad (30\ \text{mA})}$$

Answer is D.

[15] Primarily due to its excellent light yield and approximately linear response over a significant energy range.
[16] Numerous texts provide more information on scintillation theory. As a nuclear engineer, the resulting output is of more immediate concern.
[17] Semiconductor photodiodes are also used.
[18] Several types of electrical circuitry exist between the detector and the multichannel analyzer (MCA). See the Reference section for more information.
[19] Different types of detectors are available, such as lithium-drifted silicon detectors; nevertheless, the resulting output is generally the same shape and the principles are the similar. What differs is the radiation interaction in the detector itself.

[20] Both the gain and the pulse width in units of time could be taken from the manufacturer's specifications for the PM tube.

NUCLEAR INSTRUMENTS–11

One of the gammas emitted by N-16 has a wavelength of 1.75×10^{-3} Å. Assuming these are the gammas being detected by the graph shown, what are the values of points A and B?

(A) E_A 6.1 MeV E_B 2.10 MeV
(B) E_A 6.1 MeV E_B 5.84 MeV
(C) E_A 7.1 MeV E_B 3.10 MeV
(D) E_A 7.1 MeV E_B 6.84 MeV

Solution:

Point A is the photopeak or full energy peak. It represents the situation in which the energy from the incoming gamma radiation interactions is completely collected by the detector. Thus, the energy at point A correlates to the energy of the gamma itself.

$$E_A = E_\gamma = h\nu$$

h = Planck's constant
ν = frequency

The wavelength of a zero mass particle is given by the quantum relationship

$$\lambda = \frac{hc}{E}$$

Substitute the initial equation into the wavelength equation and solve for the frequency.

$$\begin{aligned}\lambda &= \frac{hc}{E}\\ &= \frac{hc}{h\nu} = \frac{c}{\nu}\\ \nu &= \frac{c}{\lambda}\end{aligned}$$

One can use the wavelength equation to solve for the energy directly or substitute the relationship between the frequency and the wavelength into the first equation.

Substitute data from Table 5.2 of App. 5A and solve for the energy.

$$\begin{aligned}E_A = E_\gamma = h\nu &= \frac{hc}{\lambda}\\ &= \frac{\left(4.136 \times 10^{-15}\ \text{eV}\cdot\text{s}\right)\left(3.00 \times 10^8\ \frac{\text{m}}{\text{s}}\right)}{\left(1.75 \times 10^{-3}\ \text{Å}\right)\left(\frac{10^{-10}\ \text{m}}{1\ \text{Å}}\right)}\\ &= \left(7.09 \times 10^6\ \text{eV}\right)\left(\frac{1\ \text{MeV}}{10^6\ \text{eV}}\right)\\ &= \boxed{7.09\ \text{MeV}\ \ (7.1\ \text{MeV})}\end{aligned}$$

Therefore, the answer is either C or D.

Point B is the compton edge. It represents the maximum electron recoil energy in a compton scattering event with the incoming gamma. The difference between the full energy or photopeak and the compton edge is given by

$$E_C = h\nu - E_{CR}\big|_{\theta=\pi} = \frac{h\nu}{1 + \frac{2h\nu}{m_e c^2}}$$

E_{CR} = electron recoil energy at a recoil angle of π
m_e = rest mass of an electron

Substitute the calculated value for E_A and the requisite data from Table 5.2 of App. 5A.

$$\begin{aligned}E_C &= \frac{h\nu}{1 + \frac{2h\nu}{m_e c^2}}\\ &= \frac{7.09\ \text{MeV}}{1 + \frac{(2)(7.09\ \text{MeV})}{\left(9.1 \times 10^{-31}\ \text{kg}\right)\left(3.00 \times 10^8\ \frac{\text{m}}{\text{s}}\right)^2 \left(\frac{1\ \text{J}}{1\ \frac{\text{kg}\cdot\text{m}^2}{\text{s}^2}}\right) \times \left(\frac{1\ \text{eV}}{1.602 \times 10^{-19}\ \text{J}}\right)\left(\frac{1\ \text{MeV}}{10^6\ \text{eV}}\right)}}\\ &= 0.246\ \text{MeV}\end{aligned}$$

Now calculate the energy at point B.

$$\begin{aligned} E_B &= E_A - E_C \\ &= 7.09\ \text{MeV} - 0.246\ \text{MeV} \\ &= \boxed{6.84\ \text{MeV}} \end{aligned}$$

Answer is D.

Notes: (1) When $h\nu \gg m_ec^2/2$ the value of E_C is

$$\begin{aligned} E_C &= \frac{h\nu}{1+\dfrac{2h\nu}{m_ec^2}} \\ &\cong \frac{m_ec^2}{2} \\ &\cong 0.256\ \text{MeV} \end{aligned}$$

Therefore, using this value for E_C can provide a quick, and relatively accurate, estimate of the energy at point B. (2) Region 7 is due to multiple compton scattering events occurring with medium-energy gamma ray input.

NUCLEAR INSTRUMENTS–12

In the detector (i.e., the scintillator) the following events can occur.

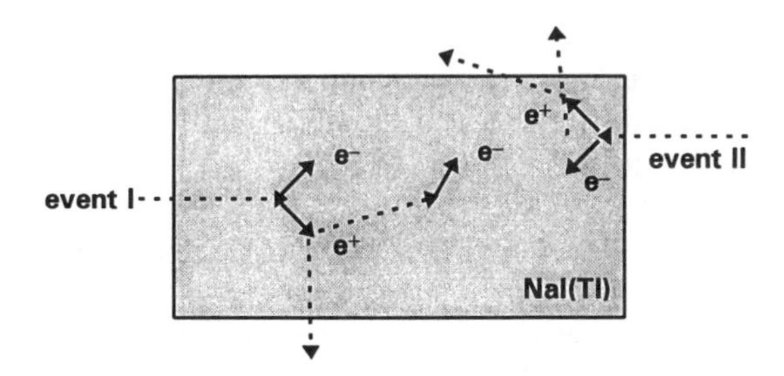

Assuming the event I gamma possesses the minimum energy necessary, how much energy is deposited in the detector? Event II represents what region on the graph?

(A) 0.256 MeV; 5
(B) 0.511 MeV; 5
(C) 0.765 MeV; 1
(D) 1.02 MeV; 1

Solution:

The gamma in event I results in pair production; therefore, the minimum energy required is 1.02 MeV. One-half that energy is carried by the positron. When the positron undergoes annihilation with an electron, two gammas are released each carrying the rest mass energy of 0.511 MeV. Since one of these escapes, that amount of energy is not detected.

$$\begin{aligned} \begin{pmatrix}\text{energy}\\ \text{deposited}\end{pmatrix} &= \begin{pmatrix}\text{incoming}\\ \text{energy}\end{pmatrix} - \begin{pmatrix}\text{energy lost}\\ \text{to the}\\ \text{detector}\end{pmatrix} \\ &= 1.02\ \text{MeV} - 0.511\ \text{MeV} \\ &= \boxed{0.511\ \text{MeV}} \end{aligned}$$

This results in a single escape peak, region 6, whose position of the graph is given by the following.

$$E_6 = E_A - m_ec^2$$

Consider event II. The loss of both annihilation photons occurs and thus causes a double escape peak located at region 5.

Answer is B.

NUCLEAR INSTRUMENTS–13

In a particular application, a scintillation detector is surrounded by lead as shown in order to measure the response of a radioactive source.

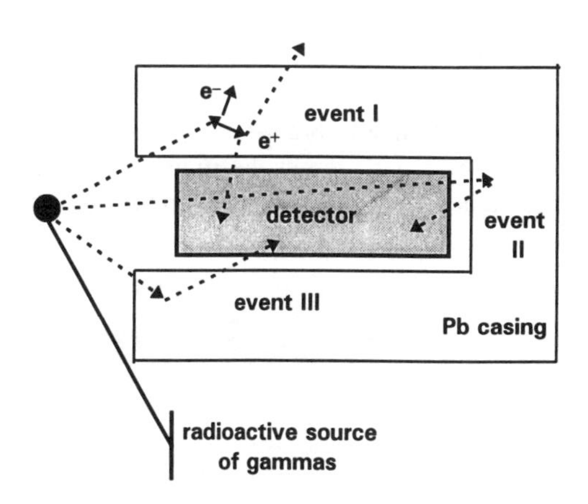

Which event is responsible for region 2 on the graph? for region 3?

(A) region 2: I region 3: III
(B) region 2: II region 3: III
(C) region 2: II region 3: I
(D) region 2: III region 3: II

Solution:

Region 2 on the graph represents peaks that occur due to the characteristic x-rays which are emitted by material surrounding the detector. In this case, the 0.074 MeV peak occurs due to the K_α x-ray from lead represented by event III.

Region 3 is the result of *backscatter* from compton scattering interactions and is shown as event II. This occurs because as the scattering angle becomes large (e.g., greater than 120°) the energy transfer to the scattered gamma becomes roughly constant regardless of the energy of the incoming gamma.[21] Also, as described in Prob. 11, when the primary (i.e., the incoming) gamma energy is much greater than one-half the rest mass energy of an electron, the associated energy is 0.256 MeV. In terms of equations, when

$$h\nu >> \frac{m_e c^2}{2}$$

$$h\nu'|_{\theta=\pi} \cong \frac{m_e c^2}{2} = 0.256\ \text{MeV}$$

$h\nu'$ = scattered gamma energy

Thus, the backscatter peak occurs between 0.20 MeV and 0.25 MeV regardless of the material.

Answer is D.

Note: Region 1 is due to bremsstrahlung radiation caused by stopping beta particles in the radioactive source and occurs in higher energy gamma spectrums.

[21] One should recall though it is the mid-range energy gammas which undergo compton scattering.

NUCLEAR INSTRUMENTS–14

Scintillation and Geiger-Mueller detector charge outputs are often large enough to be used directly in the follow on electronic circuitry. Most other types of detectors require signal amplification. For that purpose, a simple preamplifier is shown.

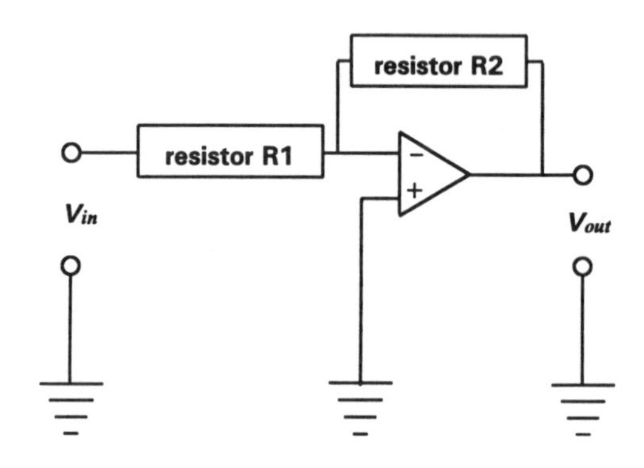

The gain of the amplifier is 10^5. The values of resistors R1 and R2 are 100 Ω and 10 kΩ respectively.

If the input voltage is 3 mV, what is the output voltage?

(A) −0.30 V
(B) −3.00 V
(C) −30.0 V
(D) −300 V

Solution:

The circuit shown is an inverting operational amplifier. Three simplifying assumptions are normally used when evaluating such amplifiers:

(1) the operational amplifier operates in the linear region,
(2) the input current (in this case the current at the negative input of the amplifier) is zero, and
(3) the voltage difference between the negative and the positive input terminals is zero.

The circuit is redrawn as shown for clarification. Call the node closest to the negative input terminal, between R1 and R2, node 1. Apply kirchoff's current law (KCL) at this node.

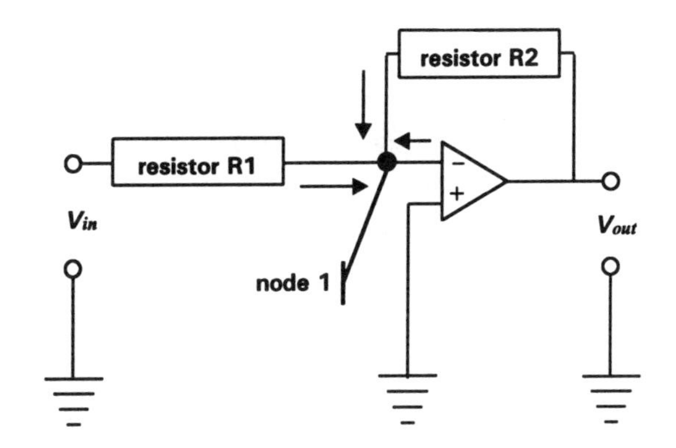

Kirchhoff's current law results in the following.

$$\frac{V_{\text{in}} - V_1}{100\,\Omega} + \frac{V_{\text{out}} - V_1}{10\,000\,\Omega} + (-i_{\text{in}}) = 0$$

$$i_{\text{in}} = \frac{V_{\text{in}} - V_1}{100\,\Omega} + \frac{V_{\text{out}} - V_1}{10\,000\,\Omega}$$

Using the assumptions means i_{in} is zero and, since the ground is attached to the positive terminal,

$$V_1 = V_- = V_+ = 0$$

Substituting this information gives the following.

$$i_{\text{in}} = \frac{V_{\text{in}} - V_1}{100\,\Omega} + \frac{V_{\text{out}} - V_1}{10\,000\,\Omega}$$

$$0 = \frac{V_{\text{in}} - 0}{100\,\Omega} + \frac{V_{\text{out}} - 0}{10\,000\,\Omega}$$

$$V_{\text{out}} = \left(\frac{10\,000\,\Omega}{100\,\Omega}\right)(-V_{\text{in}})$$

$$= -100V_{\text{in}}$$

Substitute the given value for V_{in} to obtain the desired result.

$$V_{\text{out}} = -100V_{\text{in}}$$

$$= (-100)(3\,\text{mV}) = (300\,\text{mV})\left(\frac{1\,\text{V}}{10^3\,\text{mV}}\right)$$

$$= \boxed{-0.30\,\text{V}}$$

Answer is A.[22]

[22] The gain A comes into play only in that as long as A >> R2/R1 the output is independent of the operational amplifier parameters. See the Reference section for more information.

NUCLEAR INSTRUMENTS–15

The detection efficiency for neutrons of a certain BF_3 tube with a boron density of $2 \times 10^{19}\ \text{cm}^{-3}$ is given by[23]

$$\mathit{eff}(E) = 1 - \exp(-\Sigma_a(E)L)$$

$\Sigma_a(E)$ = macroscopic absorption cross section of B-10 at energy E

L = active length of the tube

The approximate ranges in which nuclear radiation detectors used as power sensors are shown.[24]

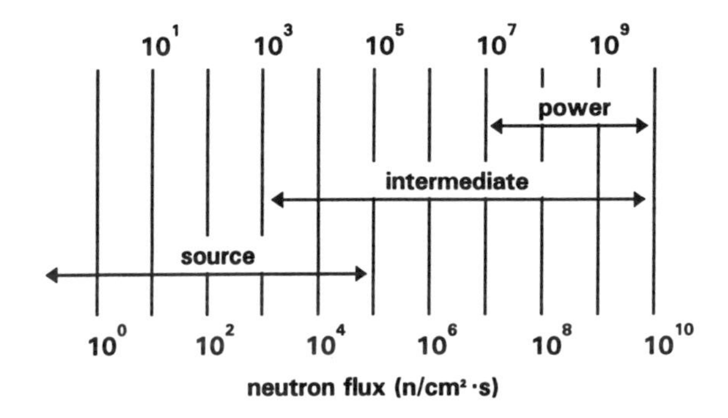

How long must the active tube length be to get at least 90% efficiency for thermal neutrons? In what range would this detector be used?

(A) 30 cm; power range
(B) 60 cm; intermediate range
(C) 30 cm; source range
(D) 60 cm; source range

Solution:

From data provided for boron in Table 5.3 of App. 5A, the microscopic absorption cross section for thermal neutrons is 3838 barns. Substitute this and the given values and solve for the desired length.

$$\mathit{eff}(E) = 1 - \exp(-\Sigma_a(E)L)$$

$$\mathit{eff}(E_0) = 1 - \exp(-N\sigma_a L)$$

$$\exp(-N\sigma_a L) = 1 - \mathit{eff}(E_0)$$

[23] This is an approximate relationship and varies with the tube pressure, construction details, and neutron angle of incidence.

[24] The flux indicated is that which exists at the detector location outside the core.

Continue by taking the natural logarithm of both sides.

$$\exp(-N\sigma_a L) = 1 - eff(E_0)$$

$$-N\sigma_a L = \ln\left(1 - eff(E_0)\right)$$

$$L = \frac{\ln\left(1 - eff(E_0)\right)}{-N\sigma_a}$$

$$= \frac{\ln(1-0.9)}{-\left(2\times 10^{19}\ \frac{\text{atoms}}{\text{cm}^3}\right)(3838\ \text{barns})\left(\frac{10^{-24}\ \text{cm}^2}{1\ \text{barn}}\right)}$$

$$= \boxed{30\ \text{cm}}$$

BF_3 detectors operate in the pulse mode and have low interaction rates with gamma radiation without external circuitry to compensate in some manner. Thus, they are not meant for high flux rates or ranges where gamma discrimination is necessary. They operate in the source range.

Answer is C.

12 NUCLEAR ISSUES: SOLUTIONS

NUCLEAR ISSUES–1

Which of the following are important issues and concerns for the nuclear industry in the future?

(A) Kyoto protocol
(B) DOE nuclear waste policy
(C) NRC licensing and decommissioning policy
(D) all of the above

Solution:

Note: Some questions on the nuclear professional engineer examination relate to current issues surrounding nuclear power. Many of these issues will not be well covered in text books due in part to the time lag from writing to publishing. Reading industry periodicals and general media sources should keep one current. Practicing engineers and others concerned with nuclear issues maintain their expertise, general knowledge level, and reputation by understanding the answers to questions such as this.

The Kyoto protocol contains a list of goals regarding greenhouse gas emissions. The protocol was generated at the United Nations Framework Convention on Climate Change in Kyoto, Japan. For the US to meet the emissions goals would require a decline of approximately 30% in the use of coal and oil. Obviously, to meet such requirements while maintaining the energy production and lifestyle America currently enjoys will require the use of an energy source with minimal greenhouse emissions and maximum practicality with minimum developmental needs—nuclear energy.

The Department of Energy was required by law to accept used nuclear fuel from industry plant sites by January 31, 1998, a deadline which was not met. Further, nuclear utility companies have paid into a Nuclear Waste Fund designed to finance the creation of such a site. The resolution of this problem is pivotal to the future of the nuclear electric power industry.

It has been a considerable time since the last order for a civilian nuclear plant in the United States. While several advanced designs which take advantage of passive safety systems to address public concerns over the use of nuclear energy, the uncertain regulatory environment continues to engender concerns about committing to the construction of a new nuclear plant. Several agencies have taken the lead—the *Institute for Nuclear Power Operations* (INPO) as a single example—in attempting to redefine the licensing process to provide for necessary public input while providing reasonable business expectations, and ultimately best serving the needs of the country as a whole. This issue has a direct impact on the future use of nuclear energy.

Lastly, until the resolution of the licensing issue mentioned, as well as the waste disposal issue, current nuclear plants will be required to operate for longer periods of time in a safe manner. How to accomplish this and effectively decommission those plants no longer required presents challenges to nuclear professionals in many fields.

In summary, all the issues mentioned are important and of concern to the nuclear industry as a whole.

Answer is D.

APPENDIX 1.A

NUCLEAR POWER SYSTEMS

TABLE 1.1
Properties of Saturated Steam by Temperature
(English Units)

		enthalpy (Btu/lbm)		
temperature (°F)	pressure (lbf/in²)	saturated liquid h_f	evaporation h_{fg}	saturated vapor h_g
440	381.2	419.0	786.3	1205.3
540	961.5	536.4	657.5	1193.8
560	1131.8	562.0	625.0	1187.0

TABLE 1.2
Properties of Saturated Steam by Pressure
(English Units)

	enthalpy (Btu/lbm)			entropy (Btu/lbm-°R)		
pressure (lbf/in²)	saturated liquid h_f	evaporation h_{fg}	saturated vapor h_g	saturated liquid s_f	evaporation s_{fg}	saturated vapor s_g
4.0	120.9	1006.4	1127.3	0.2198	1.6426	1.8624
800.0	509.7	689.6	1199.3	0.7110	0.7050	1.4160
1200.0	571.7	612.3	1183.9	0.7112	0.5961	1.3673

TABLE 1.3
Bessel Functions of the First Kind of Order 0

x	0.0	0.1	0.2	0.3	0.4	0.5	0.6	0.7	0.8	0.9	1.0
$J_0(x)$	1.0000	0.9975	0.9900	0.9776	0.9604	0.9385	0.9120	0.8812	0.8463	0.8075	0.7652

x	1.0	1.1	1.2	1.3	1.4	1.5	1.6	1.7	1.8	1.9	2.0
$J_0(x)$	0.7652	0.7196	0.6711	0.6201	0.5669	0.5118	0.4554	0.3980	0.3400	0.2818	0.2239

x	2.0	2.1	2.2	2.3	2.4	2.5	2.6	2.7	2.8	2.9	3.0
$J_0(x)$	0.2239	0.1666	0.1104	0.0555	0.0025	-0.0484	-0.0968	-0.1424	-0.1850	-0.2243	-0.2601

$$J_0(x) = 1 - \frac{x^2}{(2^2)(1!)^2} + \frac{x^4}{(2^4)(2!)^2} - \frac{x^6}{(2^6)(3!)^2} + \ldots$$

APPENDIX 1.B

NUCLEAR POWER SYSTEMS

TABLE 1.4
Properties of Saturated Steam by Temperature
(SI Units)

temperature (°C)	specific volume υ_g (m³/kg)	enthalpy (kJ/kg) saturated liquid h_f	evaporation h_{fg}	saturated vapor h_g
300	0.0216	1344.0	1404.9	2749.0
320	0.0155	1461.5	1238.6	2700.1

TABLE 1.5
Properties of Saturated Steam by Pressure
(SI Units)

pressure (kPa)	specific volume υ_f (m³/kg)	specific volume υ_g (m³/kg)	enthalpy (kJ/kg) saturated liquid h_f	evaporation h_{fg}	saturated vapor h_g
345	0.00107	0.5324	582.0	2149.6	2731.6
379	0.00108	0.4884	596.2	2139.7	2735.9
414	0.00108	0.4489	609.9	2130.2	2740.1
7000	1.3513 cm³/g	27.37 cm³/g	1267.0	1505.1	2772.1

APPENDIX 2.A

NUCLEAR FUEL MANAGEMENT

TABLE 2.1
Thermal Conductivity at 650°K for Common Reactor Plant Materials
(SI Units)

material	thermal conductivity (W/m·K)
UO_2	4.6
Zircaloy	17.0
Stainless Steel	19.0
carbon steel	41.0

TABLE 2.2
Neutron Absorption Cross Sections (0.0253 eV neutrons)

element	absorption cross section ($\times 10^{-24}$ cm^2)
Oxygen (O)	0.00027
Iron (Fe)	2.55
Zirconium (Zr)	0.185

APPENDIX 2.B

NUCLEAR FUEL MANAGEMENT

TABLE 2.3
Low Level Waste Classification Guidance

class[a]	description	requirements[b]
A	The class with the least amount of radioactivity. Waste becomes nonhazardous during the period of state or federal control following shutdown (called the institutional control period).	(1) in Table A and the concentration is ≤ 0.1 times the values shown in the table (2) in Table B and the concentration and the concentration is ≤ the value shown in column one (3) the waste contains none of the nuclides shown in Tables A or B
B	This is more restrictive than class A. Waste is hazardous up to 300 years.	(1) in Table B and the concentration is > the value shown in column one but ≤ the value shown in column two (2) marked by an asterisk (*) unless other isotopes in the low level waste (LLW) result in a more restrictive classification
C	This is more restrictive than class B. Items are still allowed to be buried in near-surface burial sites but may require greater depths than utilized for class A or B items. Waste is hazardous for greater than 300 years.	(1) in Table A and the concentration is > 0.1 times, but less than, the value shown (2) in Table B and the concentration is > the value shown in column two but ≤ the value shown in column three
GTCC	This class takes its name from Greater Than Class C. Waste is not suitable for shallow burial.	(1) in Table A and > the value shown (2) in Table B and > the value shown in column three

[a] The most restrictive classification applies.
[b] If a mixture of radionuclides is present, the sum of fractions of the individual isotope concentrations as compared to a particular column should be used to determine the classification. A sum of less than 1.0 allows the waste to be classified by that column's limit. See the Nuclear Fuel Management section for an example.

Note: Table 2.3 is a partial table for instruction purposes only. The Table A and Table B referred to are the example tables used in the text problem in the Nuclear Fuel Management section. Though a partial table, the text questions have been written so that classification using this table is consistent with that which would be obtained with the complete regulations in 10CFR61

TABLE 2.4
Half-lives of Several Nuclides of Concern

nuclide	half-life ($T_{1/2}$)
Cs-137	30.17 yr
Pu-241	14.4 yr
I-129	1.57×10^7 yr
Ni-59	7.6×10^4 yr
Ni-63	100 yr
P-32	14.28 days
Pu-240	6.56×10^3 yr
Pu-241	14.4 yr

APPENDIX 3.A

NUCLEAR RADIATION

TABLE 3.1
Rest Mass of Elementary Particles

particle	mass
m_e	9.109×10^{-31} kg
m_n	1.675×10^{-27} kg
m_p	1.673×10^{-27} kg

TABLE 3.2
Mass Attenuation Coefficients for Selected Materials

material	γ energy (MeV)	μ/ρ (cm^2/g)
Pb	1.33	5.67×10^{-2}
Pb	1.00	6.84×10^{-2}
ordinary concrete	1.00	6.495×10^{-2}

TABLE 3.3
Mass Absorption Coefficients for Selected Materials

material	γ energy (MeV)	μ_a/ρ (cm^2/g)
Fe	1.00	2.603×10^{-2}
Si	1.00	2.778×10^{-2}
body tissue[a]	1.00	3.108×10^{-2}

[a] The value of the mass absorption coefficient for body tissue is very close to that of water.

TABLE 3.4
Data on Selected Radionuclides

material	half-life	modes of decay	gamma decay energy (MeV)
$^{56}_{25}Mn$	2.58 hr	β^-, γ	0.8; 1.8; 2.1
$^{59}_{26}Fe$	44.51 days	β^-, γ	1.1; 1.3
$^{16}_{7}N$	7.13 s	β^-, γ	6.1; 7.1
$^{41}_{18}Ar$	1.82 hr	β^-, γ	1.3
$^{135}_{55}Xe$	9.10 hr	β^-, γ	0.3

APPENDIX 3.B

NUCLEAR RADIATION

TABLE 3.5
Point Buildup Factor Parameters:
Taylor Form
$B_p = A_1\exp(-\alpha_1 \mu r) + A_2\exp(-\alpha_2 \mu r)$

material	energy (MeV)	A_1[a]	$-\alpha_1$	α_2
Pb	0.5	1.677	0.03084	0.30941
Pb	1.0	2.984	0.03503	0.13486
Pb	2.0	5.421	0.03482	0.04379

[a] Since as r approaches zero, the buildup factor B_p approaches unity (there can be no buildup factor without a shield), $A_1 + A_2 = 1$ and, therefore, only A_1 need be tabulated.

TABLE 3.6
Selected EPA Standards for Radioactivity in Community Drinking Water Systems
(40CFR141; 1976)

element	concentration limit	dose limit[a]
Ra-226 plus Ra-228	0.2 Bq/L	0.04 Sv

[a] The limit is for the whole body or for any individual organ.

APPENDIX 4.A

NUCLEAR THEORY

TABLE 4.1
Selected Constants of Concern in Nuclear Engineering

item	symbol	value
atomic mass unit[a]	amu	1.66054×10^{-24} g
		931.481 MeV
		$1/N_A$ g
Avogadro's number	N_A	0.602217×10^{24} $(\text{g}\cdot\text{mol})^{-1}$
neutron rest mass	m_n	1.67492×10^{-24} g
		1.00866 amu
proton rest mass	m_p	1.67262×10^{-24} g
		1.00728 amu
hydrogen atom rest mass	m_H	1.67353×10^{-24} g
		1.007825 amu
Boltzmann's constant	k	1.38×10^{-23} J/K
speed of light	c	3.00×10^8 m/s

[a] The amu is defined as one-twelfth the mass of a neutral C-12 atom.

TABLE 4.2
Selected One-Group Constants: Fast Reactor[a]

material	σ_f (barns)	σ_a (barns)	ν	η	σ_{tr} (barns)
Fe-55	—	0.006	—	—	2.7
U-235	1.4	1.65	2.6	2.2	6.8
Pu-239	1.85	2.11	3.0	2.6	6.8

[a] One-group calculations are most applicable to fast reactors. Two-group calculations, as a minimum, are required for thermal reactors.

TABLE 4.3
Typical Values for Nuclear Parameters: Thermal Reactor
(U-235 at 20°C)

parameter	value
η_{th}	2.065
ρ	0.95
ε	1.05
g_a	0.9780
g_f	0.9759
σ_a	681 barns
σ_f	585 barns
$\sigma_{a,\text{H2O}}$	0.664 barns

APPENDIX 4.B

NUCLEAR THEORY

TABLE 4.4
Selected Diffusion Parameters for Water
(ρ = 1.0 g/cm^3 at 20°C)

item	symbol	value
thermal diffusion area	L_T^2	8.1 cm^2
thermal diffusion coefficient	$\overline{D}$	0.16 cm
fast diffusion coefficient	D_1	1.13 cm
neutron age	τ_T	27 cm^2

TABLE 4.5
Selected Delayed Neutron Fractions β for Thermal Neutron Fission

material	β
U-233	0.0026
U-235	0.0065
Pu-239	0.0021

APPENDIX 5.A

NUCLEAR INSTRUMENTS

TABLE 5.1
Electromagnetic Spectrum

designation	frequency (Hz)	wavelength (Å unless otherwise noted)	approximate photon energy (eV unless otherwise noted)
radio	$1 \times 10^5 - 3 \times 10^{10}$	3×10^5 cm – 1.0 cm	$4 \times 10^{-10} - 1 \times 10^{-4}$
infrared	$3 \times 10^{12} - 3 \times 10^{14}$	0.01 cm – 10,000	0.0124 – 1.24
visible	$4.3 \times 10^{14} - 3 \times 10^{16}$	7000 – 100	1.77 – 124
x-ray	$3 \times 10^{16} - 3 \times 10^{23}$	100 – 0.00001	124 –1240 MeV
γ-ray	$3 \times 10^{18} - 3 \times 10^{21}$	1 – 0.001	12.4 keV – 12.4 MeV

TABLE 5.2
Selected Constants and Conversions

quantity	symbol	value
Planck's constant	h	6.6256×10^{-34} J·s
		4.136×10^{-15} eV·s
speed of light	c	3.00×10^8 m/s
angstrom	Å	10^{-10} m/Å
electron rest mass	m_e	9.109×10^{-31} kg
joules to eV	J and eV	1.602×10^{-19} J/eV

TABLE 5.3
Thermal Neutron Absorption Cross Sections

material	microscopic cross section (barns)
${}^{10}_{5}B$	3838
${}^{3}_{2}He$	5330
${}^{16}_{8}O$	0.19×10^{-3}

REFERENCE LIST

BOOKS

American Nuclear Society. *Glossary of Terms in Nuclear Science and Technology.* La Grange Park, IL: American Nuclear Society, 1986.

Beiser, Arthur. *Concepts of Modern Physics.* 3rd ed. New York: McGraw-Hill, 1981.

Blatt, John M., and Victor F. Weisskopf. *Theoretical Nuclear Physics.* 1952. Reprint, New York: Dover Publications, 1989.

Bohm, David. *Quantum Theory.* 1951. Reprint, New York: Dover Publications, 1989.

Cochran, Robert G., and Nicholas Tsoulfanidis. *The Nuclear Fuel Cycle: Analysis and Management.* La Grange Park, IL: American Nuclear Society, 1990.

El-Wakil, M. M. *Nuclear Heat Transport.* La Grange Park, IL: American Nuclear Society, 1978.

Etherington, Harold, ed. *Nuclear Engineering Handbook.* New York: McGraw-Hill, 1958.

Glasstone, Samuel, and Alexander Sesonke. *Nuclear Reactor Engineering.* 2 vols. New York: Chapman and Hall, 1994.

Keenan, Joseph H., and others. *Steam Tables.* New York: John Wiley and Sons, 1969.

Keryszig, Erwin. *Advanced Engineering Mathematics.* 4th ed. New York: John Wiley and Sons, 1979.

Knoll, Glenn F. *Radiation Detection and Measurement.* 2nd ed. New York: John Wiley and Sons, 1989

LaMarsh, John R. *Introduction to Nuclear Engineering.* 2nd ed. Reading, MA: Addison-Wesley, 1983

Larsen, Richard J., and Mossis L. Marx. *An Introduction to Mathematical Statistics and its Applications.* Englewood Cliffs, NJ: Prentice-Hall, 1981.

Lindeburg, Michael R. *Engineer-in-Training Reference Manual.* 8th ed. Belmont, CA: Professional Publications, 1998.

———. "Nuclear Engineering." Chap. 18 in *Mechanical Engineering Reference Manual.* 8th ed. Belmont, CA: Professional Publications, 1990.

———. *Engineering Economic Analysis.* Belmont, CA: Professional Publications, 1993.

Parker, Sybil P., ed. *McGraw-Hill Dictionary of Scientific and Technical Terms.* 5th ed. New York: McGraw-Hill, 1994.

Shackelford, James F.; William Alexander; and Jun S. Park, eds. *CRC Materials Science and Engineering Handbook.* 2nd ed. Boca Raton, FL: 1994.

Shleien, Bernard; Lester A. Slaback, Jr.; and Brian Kent Birky, eds. *Handbook of Health Physics and Radiological Health.* 3rd ed. Baltimore: Williams and Wilkins, 1998.

Thomas, Brian J. *The Internet for Scientists and Engineers.* 1996 ed. Bellingham, WA: SPIE Optical Engineering Press; Piscataway, NJ: IEEE Press, 1996.

WORLD WIDE WEB SITES

American Nuclear Society   http://www.ans.org/

National Council of Examiners for Engineering and Surveying   http://www.ncess.org/

Nuclear Engineering Library (U.C. Berkeley)   http://neutrino.nuc.berkeley.edu/NEadm.html

U.S. Nuclear Regulatory Commission   http://www.nrc.gov/

ABOUT THE AUTHOR

J. A. Camara is a Lieutenant Commander in the US Navy submarine service. He has over 20 years of experience in the nuclear field as a technician, watch supervisor, plant supervisor, instructor, inspector, and nuclear engineer. This experience has included engineering duty during the construction of a nuclear reactor plant, a refueling overhaul and major radiological evolutions, and soon will include the decommissioning of a reactor. He has been an instructor at the navy's nuclear power school for both enlisted and officer students, teaching physics and electrical engineering. While there, he was recognized as the Instructor of the Year. He possesses a PE license in the fields of electrical engineering and nuclear engineering. He obtained his MS in space systems from the Florida Institute of Technology and his BS in electrical and computer engineering/materials science and engineering from the University of California, Davis. He is currently progressing toward a nuclear engineering PhD at the Georgia Institute of Technology.

NO POSTAGE
NECESSARY
IF MAILED
IN THE
UNITED STATES

BUSINESS REPLY MAIL

FIRST CLASS MAIL PERMIT NO. 33 BELMONT, CA

POSTAGE WILL BE PAID BY ADDRESSEE

PROFESSIONAL PUBLICATIONS INC
1250 FIFTH AVE
BELMONT CA 94002-9979

NO POSTAGE
NECESSARY
IF MAILED
IN THE
UNITED STATES

BUSINESS REPLY MAIL

FIRST CLASS MAIL PERMIT NO. 33 BELMONT, CA

POSTAGE WILL BE PAID BY ADDRESSEE

PROFESSIONAL PUBLICATIONS INC
1250 FIFTH AVE
BELMONT CA 94002-9979

NO POSTAGE
NECESSARY
IF MAILED
IN THE
UNITED STATES

BUSINESS REPLY MAIL

FIRST CLASS MAIL PERMIT NO. 33 BELMONT, CA

POSTAGE WILL BE PAID BY ADDRESSEE

PROFESSIONAL PUBLICATIONS INC
1250 FIFTH AVE
BELMONT CA 94002-9979